To Be a Scientist

To Shirley, David,
Peter, and Jenny

Foreword

By Sir John Fairclough

It is always difficult to buck the trend: it takes courage. That is true in any aspect of life. It is certainly true in science. The peer review system can make it particularly difficult for the researcher who needs funding for an unconventional line of research. The merits of the case are often decided by the established experts in the field and, human nature being what it is, they may dismiss the application out of hand, or subject the researcher to intense cross-examination. This is a valuable and essential process—most of the time. The evaluation, of course, is aimed at ensuring that only the best research projects are supported. But the judgement of what is the best, when money is short, is likely to be biased towards research that seeks to enhance existing and accepted theories.

If the applicant has conviction and courage, and has gained some support for his or her ideas, he or she may get the funding. But what happens to the applicant whose request is denied? He or she may be fortunate and have a sponsor in their local university, or a good relationship with an enlightened company. Word gets about, however, that the application has been rejected by the peer review and this can act as a negative stigma and reduce the probability of support.

Don Braben recognized this problem and persuaded British Petroleum in the early 1980s to introduce a 'Venture Research' programme. This was to provide money for a few researchers whose record of achievement was well established but whose radical ideas, often best described as enlightened insight with a smattering of inspired guesswork, had not won the support of any funding authority. An examination of the relevant history shows that many of the theories we now take for granted were generated by people with courage and conviction who ran the gauntlet of their peers, committed to other views, and in so doing risked their reputations.

This book sets out to show why such a system of support is important—indeed mandatory—if we are to get the best from our scientific endeavours. The chosen few would have a clear vision and know that success would bring enormous value to industry or society or both. 'Picking them was easy'—according to Don Braben—but the proof is in the eating and there is ample evidence to show that he picked some real winners. It is important to be clear that he picked 'winning people' and did not pursue a policy of 'picking winners' as the term is usually applied. This is an important distinction.

This book is not just about scientists for scientists, although every researcher will find something with which he or she will identify. It will be enjoyed by the many people who admire courage and a commitment to succeed against adversity. It is a series of adventure stories and will be enjoyed as such by all. Perhaps the programme that Don Braben began could be better called 'Adventure Research', for the stories told are of people engaged in great adventures, not of a physical nature, but seeking to find something new and useful through intellectual effort. Such work is just as demanding of courage and tenacity as any of the more popular adventure seeking activities.

This book should be required reading for all those who have a responsibility to make decisions about the funding of research. The British Petroleum 'Venture Research' programme is no longer operating. It was a casualty of the need to cut costs and expenses to bolster profits during the economic downturn of the early 1990s. There is a need for a funding mechanism of modest proportions, based on the principle of picking winning people, organized separately from the peer review process to support a few able scientists.

I hope, too, that this book will give encouragement to all those scientists who are struggling to find acceptance for ideas that are rejected by the scientific establishment. To them I say: keep the faith!

Winchester, July 1993

Preface

This book is intended to provoke the curiosity of those people who have an interest, passing or professional, in the pursuit of science. Most general books on this subject are devoted to such broad scientific concepts as time, or evolution, or are concerned with the history or philosophy of science. Although I have had my conceptual, historical, or philosophical moments in writing these pages, my main purpose has been to offer a commentary on the state of being a scientist, that is on how one becomes a scientist, what scientists do, and why what they do and how they do it has importance for everyone.

The term 'scientist' is usually reserved for those professionals who make their living from scientific enterprise, but as conventionally defined (see for example the *Concise Oxford Dictionary*), the term refers to a person with expert knowledge of a science and, in turn, science is described as systematic and formulated knowledge. Although it would stretch the definition beyond its normal bounds, it could be said that any person who thinks him- or herself expert, and takes a serious approach to what he or she does, whatever it is, could be entitled to call him- or herself a scientist if he or she wished to do so. The point is not simply pedantic. One of the themes of the book is that future economic growth and the welfare of mankind will depend on the extent to which the constituency of science can be expanded far beyond the elite of its professionals. There seem to be no reasons why science should not be enjoyed, and scientists applauded or criticized by a wider range of *cognoscenti* than at present, without devaluing or diminishing what professional scientists do. On the contrary, such other intellectual pursuits as music or literature have long benefited from having an extensive, well-informed, and appreciative following.

The book paints a thumb-nail sketch of the scientific enterprise from a broad perspective that includes the natural sciences, engineering, and technology. It is intended for people of all ages, for scientists and policy-makers, and for anyone who has an interest in science or takes pleasure in matters intellectual. The sketch will be more of a cartoon than a portrait. Cartoons are supposed to be enjoyable, and this art form is perhaps more appropriate to a subject that is still in its infancy, with its great work—to provide the means by which mankind can survive the growing hazards from such threats as the push of population, starvation, disease, environmental pollution, and global warming—still to be accomplished. Scientists have already achieved a great deal, and there are signs that science is capable of meeting its biggest challenge. Just as for people, however, growing up is never easy. Science, still a youthful subject, is having to adjust to not always getting its own way, and with having to make do with much less money than scientists think it deserves. As this is a situation that most of us have to endure, it need not necessarily be a bad thing. Most of us learn to adjust, and our characters are supposed to benefit from the process. But there can be problems, and we should not be too surprised to find that the scientific enterprise has acquired some bad habits and is heading for trouble.

When a youngster goes astray, there will often be a friend or a relative who will try to excuse the behaviour, to explain that he or she might not be entirely to blame, and to point out that the *enfant terrible* will do great things given encouragement and help. My cartoon will be portrayed from a similar perspective. Although it will be drawn with affection, there will need to be some harsh lines if the cartoon is to have any effect, but I hope that they will be seen to have been tempered by good-natured humour.

The book is organized as follows. Chapter 1 gives a series of snapshots of the development of science over the last few centuries. It need hardly be said that it is not offered as a history. My intention here is simply to indicate how the practice of science has evolved into its present highly fragmented state. Chapters 2 to 5 are concerned with what scientists do; how one enters the profession, selects a problem to work on, attracts research funds, gets the results published, is promoted, secures a tenured position, wins prizes, and makes major discoveries. The emphasis in these few chapters is on academic scientists, with a little about their cousins who work in 'not-for-profit' organizations, such as government laboratories. Chapter 6, however, focuses on the other great scientific domain, and depicts some of the main features of an industrial scientist's career. The story in these chapters is told from the point of view of the individual scientist, but the final two chapters are written on a broader canvas, and I have tried to show why science is important to the well-being of society as a whole. Thus, Chapter 7 describes something of the links between science and economic growth, and the final chapter is concerned with the contribution that scientists can make to help mankind survive and prosper.

As my story develops, it is illustrated by outlines of some of the great discoveries and achievements in science and technology as we see them today. My intention, however, has been to try to capture a little of the excitement that a pioneering scientist might feel, or to say something of the impact of a work rather than to give a selection of rigorously scientific synopses. Generalizations are always dangerous. We may say, for example, that the Earth's shape is spherical, but strictly speaking we would, of course, be wrong to do so, and our error might be large enough to cause a Moon-shot to miss its target. With similar accuracy, we might tell a visiting space-traveller that all the animals on Earth were insects, because there are at least six million species of insects here compared with only four thousand of mammals, nine thousand of birds, and twenty-one thousand of fish. Indeed, one of the routes to new science is to show that an accepted generalization is wrong. In drawing my cartoons, therefore, I hope that the experts will agree that the slight imperfections this art form must sometimes contain are not out of character, and that they might help to stimulate serious and sustained interest among a wider community than their own.

Another of the book's themes is on the value of diversity. One of the characteristics of great discoveries is in the manner of their arrival, in that they often come out of the blue from an unexpected quarter. Albert Einstein's genius first flowered while he was working as a Technical Expert (Third Class) at the Swiss Patent Office in Berne; the transistor, and the dying echoes of the Big Bang, were discovered at an industrial research laboratory, as many other breakthroughs have been in the past. Nowadays, however, research is usually administered in ways that make life very difficult for scientists who work at places not renowned for their scientific excellence, or who have highly unorthodox ideas, and, furthermore, industrial research laboratories now concentrate almost entirely on problems directly related to their business interests that are expected to show a return in the short term. Thus, collectively speaking, the scientific enterprise is behaving as if most, if not all, the possible *major* discoveries have already been made, and diversity has largely had its day. As a result, there seems to be a widespread belief among research funding organizations that dissent can safely be ignored.

There is hardly a field of human endeavour, however, where we are not faced with apparently insoluble problems of enormous proportions. Many of them are seen as being predominantly social, but science is inextricably woven into our social fabric. Perhaps the greatest contribution that scientific enterprise has yet made has been to provide the means by which clean and reliable water supplies can be enjoyed by almost everyone in the advanced countries. Indeed, the state of the water supply can generally be used to decide whether a country can be considered to be advanced or not. It will be astonishing if the daunting problems that face mankind today are not alleviated by future scientific advances, but at present it is certainly not easy to see what they might be.

The point of view expressed in the following pages is that although these global problems do not seem to be susceptible to direct attack, they will become easier as we learn to extend our understanding into those areas where our ignorance has not yet been recognized. This intellectual feat is not the logical nightmare it might seem, but it will be accomplished only when we appreciate the value of diversity. I do not mean to imply that diversity was highly valued in the past. It might not have been, but at least it was not systematically suppressed on the spurious grounds of economy and efficiency, as is generally the case today.

Those who criticize the way things are have a duty to say how they might be changed. In what follows, I have from time to time drawn on my own experience in launching an initiative specifically aimed at fostering diversity. It is based on the provision of freedom and funds for those few scientists who would radically and credibly challenge conventional wisdom, or would wish to do something that consensus opinion would regard as unimportant or irrelevant. The work of such pioneers in the past has often led to new and unpredictable insight, and in turn to new and exciting industrial opportunity. In recognition of these simple facts, our initiative also from the outset develops trusting relationships between the pioneers and their imaginative counterparts in industry, while protecting the freedom of action of all concerned.

The initiative, known formally as Venture Research, or more colloquially as the Blue Skies Project, has been very successful in terms of the science it has spawned and the new types of technological potential it has created. However, the caution and introversion arising from the recession of the early 1990s not only brought the first phase of the initiative to a premature end, but has made it difficult to find new sponsors. As a result, a large part of the potential remains unrealized.

The hiatus creates a dilemma, in that if I were to try to anticipate exciting results to emerge from the Venture Research programmes I would almost certainly fail to identify the ones that will have the most industrial value. Imagine writing about the transistor a few years after its discovery. It would have been impossible then to predict the revolutionary impact the work would have: even its discoverers were taken aback by the magnitude of its eventual success. Some Venture Research programmes might have a similar impact, but it would be presumptuous to make such claims at present. In choosing examples of this type of research, I have selected those which seem best to illustrate the point being made at the time, independently of any profitable potential the research might have. I have also described my own excitement at the beginning of a crusade as the scientists involved have progressively opened a door to reveal a new, surprising, and tantalizing panorama, but for the present I have stopped short of relating many of the subequent adventures and discoveries that are still unfolding, simply because at this stage of the initiative there is so much that is changing, and books take many months to emerge! Thus, by focusing on the attractions of the *idea* of unbounded exploration rather than on how specific and predictable wonders might be found, I hope that more people will be encouraged to think about what they really want to do, and to plan their own voyages of discovery, wherever they may lead.

In my travels over the scientific landscape I have come across many people wistfully wishing that they were setting out on their scientific careers at the beginning of this century when there was so much to be discovered. Today, there is a popular myth that there are no frontiers left (except perhaps those in space, which for the foreseeable future will be accessible only to a privileged few at best). For those of us who see attractions in new frontiers, however, the good news is that they will be found in abundance if we will go out of our way to look for them. That they exist in profusion can be evidenced from the paucity of our current *understanding* despite a profusion of information in almost every field of human endeavour. It is likely that a hundred years from now scientists will be yearning for the virgin fields of the 1990s. The bad news is that they are likely to be found in lonely places, and bandwagons never go there.

From the industrialist's point of view, new frontiers almost invariably open the way to profitable opportunity. As will be discussed in Chapter 7, one of the causes of the grim recession of the early 1990s might have been overcrowding in the current technological mainstreams. As new entrants come along, it becomes more difficult to maintain competitiveness, product ranges shrink, development costs go up, margins are driven down, and flexibility of response falls, eventualities which will all increase as a technology ages. By fostering diversity industrialists would not only be helping themselves; they might also help to provide precious keys for global survival.

Essex   D.W.B.
May 1993

Acknowledgements

In common with many other pursuits, science is a highly social activity, and the depth and extent of one's comprehension and the degree of enthusiasm for one's work is often critically dependent on the state of one's relationships with friends and colleagues. In my experience, open admissions of ignorance accompanied by calls for help are rarely ignored, and as I can claim to be expert in that sorry state of mind, I have benefited from the insight of many hundreds of scientists over the years, and I am immensely grateful to them all. I owe a special debt to those who have given their help, and have commented on the various stages of this little book, and particularly to Steve Baker, Jocelyn Bell Burnell, Mike Bennett and Pat Heslop-Harrison, Keith Briffa, Steve Davies, Nigel Franks, Netty van Gasteren, Paul Geroski, David Graham, Dudley Herschbach, Andrew Hodges, Jeff Kimble, Peter Noble, Graham Parkhouse, Ian Ross, Tony Regnier, Tim Sanderson, Colin Self, Harry Swinney, and Robin Tucker, and to Edsger W. Dijkstra, who has assiduously strived to reduce the worst excesses of my abuse of the English language. They have all been unstinting in their help, but of course, I take full responsibility for the errors and omissions that probably remain.

I am grateful for the encouragement of the Oxford University Press in helping me bring this book to fruition. I would also like to thank the Principal of Queen Mary and Westfield College London, Professor G.J. Zellick, for permission to use the university library, and to the very helpful staff there.

My parents, Agnes and Walter Braben, have been a constant source of encouragement and wisdom over the years, and I am grateful for their prodigious efforts in ensuring, when times were sometimes very hard, that I received the best education Liverpool had to offer. The youthful zest and exuberance of my children, David, Peter, and Jenny have been a great stimulant, especially when they treat me as an equal, or I take myself too seriously.

Without the sustained commitment, support, and devotion of my wife Shirley, however, and her tireless labours in typing, correcting, cajoling, and generally helping to manage the mountains of paper that threaten to engulf an author, I doubt whether this book would have been finished.

Contents

1

The scientific landscape

Scientists are a curious band of people who have an insatiable craving for knowledge. Not just any knowledge. It must be new if they are to get some relief from their condition, but, of course, it can only be temporary as today's discoveries soon become yesterday's received wisdom, at which point the craving returns. Journalists have a similar problem, but for the scientist everything published must be shown to be a scoop on a global scale; an exclusive report on some new aspect of the ways Nature runs things.

The need for exclusivity means that scientists not only have to be good at interrogating Nature, but they also have to be reasonably sure that no-one else is asking similar questions, or confident that they will get the answers before the competition does. In principle, therefore, all scientists should be aware of what has been done, and of what is going on in every other laboratory or place of learning in the world if they are to avoid the risk of coming second, or worse still, of reinventing the wheel.

This was certainly possible during the centuries up to the Renaissance. Scientists then were generally aware of each other's work because they were so few. Thereafter, however, the joys of unrestrained enquiry after more than a thousand years of intellectual repression attracted so many recruits that by the eighteenth century it had become impossible for an individual to be aware of more than a portion of what was going on. Up to that time if you were contemplating becoming a scientist you could survey the whole of science, or natural philosophy as it was then called, and turn your attention to any area that took your fancy. The legendary Isaac Newton, for example, is best remembered for the discovery published in 1686 that the force which eventually brings a cannon-ball to earth is exactly the same as that which holds the moon and the planets in their orbits. This seminal discovery rocked the world, because it embraced so many apparently unrelated phenomena: the tides, comets, stars, pendulums, and of course the apocryphal falling apple. It was indeed the first recognition of universality in

Nature, and just as important, it opened our minds fully to the vastness and complexity of the universe we live in.

The heavens have been studied for as long as we have had eyes to see, and quite literally that is how it was done until the discovery of the telescope in about 1600. The monumental achievement of mapping the planets and stars carried out in the sixteenth century by Tycho Brahe, later with the assistance of Johannes Kepler, was achieved with nothing more than their aching eyes. But it was Kepler who realized that their colossal collection of data could be summarized by a few succinct equations describing, for example, how the planets move in ellipses rather than circles, and in general laid the foundations for Newton's great work.

It was not until 1610 that Galileo Galilei reported the results of turning his telescope skywards with what must have been trembling anticipation. Although he did not invent the telescope (indeed, there is disagreement over who did), he was probably the first person to use one effectively. He saw stars in profusion which no-one had ever recorded, and he could see that the planets really were other worlds like our own. His telescope had a magnification of about 1000, and perhaps to his surprise, it was unable to magnify stellar images except to make them brighter. He guessed, therefore, that they must be very far away, but he was unable to prove his conjecture or to put a scale on their distance. Nevertheless, his punchy pamphlet, published in Latin of course, had a dramatic effect, and landed him in very deep waters with influential academics and the suffocating parochialism of their belief that placed the Earth and hence themselves at the centre of a stunted, subservient universe. It subsequently, alas, led to tragic confrontation with the Church.

Newton, however, was able to calculate that since the stars exert no noticeable gravitational pull on the planets, they must at least be hundreds of times as distant as Saturn, then the furthest planet known. Stars, therefore, were too far away to shine on us by light reflected from the Sun as the planets do, and must generate their own light, as the Sun does. As suns in their own right, each star might therefore have its own planetary system, and perhaps even its own Earth. Not since the Garden of Eden had an apple awakened such an avalanche of awareness.

Today, it is hardly possible to imagine the impact of Newton's heretical ideas. Not surprisingly, they were intensely controversial, and not only for their religious implications. Newton could explain the net result of the effects of gravitation, by calculations on the Moon's orbit for example, but much to his own dissatisfaction he could not explain to his many critics *how* the result came about: how, for example, the gravitational forces between the Sun and the Earth make themselves felt apparently instantaneously across the void of space. This problem was neatly solved by Einstein, but in spite of the enormous progress since Newton's day, the agent that transmits changes in the gravitational force has still not been observed. We are much more familiar with Nature's apparently more exotic forces—which, by the way, go under the names of electromagnetic, strong, and weak—than with some aspects of the force of gravity which plays such an important and obvious role in our everyday lives.

However, even if Isaac Newton had done none of this, his fame would have spread across the centuries for any one of a wide range of major discoveries that many scientists today would give their eye-teeth to make. In mathematics he developed the differential calculus for his work on gravitation, which is now an essential tool for almost every practising scientist. In mechanics, his laws of motion, the ubiquitous 'Newton's laws', turned much of engineering from an art into a science. He carried out extensive researches in optics, and among many other things showed that white light was a composite of all the colours of the rainbow. He is usually credited with inventing the reflecting telescope, and the influence of one of his designs can still be seen in the most advanced astronomical telescopes in use today. Indeed, it is remarkable that so much of his work is still relevant.

But he did not quite have the Midas touch. Although it is barely credible, Newton spent a large part of his life in pursuit of a Holy Grail that even the most desperate and ideas-starved scientist today would not contemplate for a second, the general outcome of which is about as relevant to today's world as witchcraft. Nevertheless, over a period of some thirty years, Newton dedicated himself to an extensive series of studies and experimentation in alchemy. It seems to have been fruitless. One can only assume that he was unable to free his mind from the rigid dogma prevailing at the time that every substance, animate or inanimate, was composed in varying proportions of just four elements—earth, air, fire, and water—and that there might therefore exist some primary substance—the philosopher's stone—from which every other substance, including gold, might be derived.

For over two thousand years, from about the fifth century BC until it gradually faded away during the eighteenth century, this restrictive concept dominated thinking in every culture and every place of learning. Many of Newton's voluminous works on alchemy have never been published, but the British economist, John Maynard Keynes acquired a large collection of Newton's manuscripts at a Sotheby auction in 1936. Keynes's conclusion was that Newton 'was the last of the magicians . . .' and that his alchemical work was 'wholly devoid of scientific value'. But Isaac Newton was one of the greatest men who has ever lived, and it is both astonishing and worrying that even an intellect of such immense proportions as his was unable to cut through the brainwashing fog of mysticism that held back progress in what we now call chemistry for so long. It is also a great pity that this extended episode is so little discussed today, as it has lessons for us all.

Towering genius though Newton was, the vast range of his interests was not unique. Even during the eighteenth century there were many great scientists—Daniel Bernoulli, Henry Cavendish, Leonard Euler, Benjamin Franklin, and Edward Halley among many others—who each turned their attention to a wide range of subjects whose scope would be virtually impossible for an individual to encompass today. Progressively throughout the nineteenth century, and rapidly since then, the expansion of knowledge has been so great that scientists have been forced to choose the part of the vast panorama that best suits their temperament and abilities. Initially, during the nineteenth century, the choices were between a relatively small number of largely independent disciplinary fields such as biology, chemistry, mathematics, medicine, and physics, which together made up the whole of what was natural philosophy. Fields require boundaries of course, and these became established during the century as large numbers of new universities were built and autonomous departments specializing in each field were created. These broad structures persist to the present day, but they have each been further subdivided many times as scientific enterprise has exploded in importance and scope.

Over a period of two hundred to three hundred years, therefore, natural philosophy, or science as it would be called today, has moved in importance from being the name used to describe the perennial preoccupation of its practitioners to a largely meaningless catch-all term. Some years ago, travelling between Canada and the USA, I was asked by a zealous customs official about my occupation. When I said I was a scientist she became suspicious and told me quite bluntly that she did not believe me. 'Everyone knows you have to specialize,' she said, and held me up some time before she was convinced that I was a suitable person to allow into her country. She was quite right, as hardly any scientists practising nowadays would refer to themselves by that title; they would use terms like clinical biochemist, plant molecular biologist, inorganic chemist, condensed matter physicist, or one of a few hundred other labels to describe what they do. Furthermore, any one of these apparently narrowly defined fields today would cover a greater volume of activity than the whole of science or natural philosophy did in Newton's time.

Not surprisingly, the expansion of science has also been accompanied by fission, as each of the main disciplinary continents has, so to speak, separated from the supercontinent of natural philosophy in much the same way as the Earth's continents drifted away from the gigantic land mass of Pangaea some 200 million years ago. The gradual separation of the main disciplines from the parent and from each other has also led to the development of different languages, cultures, and customs for each of them. The terrains of, say, the Americas and Africa are of course different, and the differences can be fascinating. Their seemingly endless variety has inspired countless explorers over the centuries in their quests to be the first to bring news of yet another remarkable landscape lying just over the horizon. But atlases of facts do not necessarily make for understanding, and although one can be a successful explorer in the Americas while knowing nothing of Africa, it was inevitable that from time to time scientists would hear news of discoveries from what they thought were quite different lands to remind them of Nature's affection for universality.

The life sciences, for example, have been dominated up to the last few decades by the qualitative, descriptive approach, that is, by the need to classify and name the millions of types of living species resident on our planet, and to describe their performance at every level from the smallest part of their constituent cells to the individual and collective behaviour patterns characteristic of the species.

For the physical sciences, experimentation and rigorously quantified measurement have been important since Newton's time, although it is curious that chemistry had its feet in both camps even as late as the 1930s when the qualitative, non-numerical attributes of the subject were still recognized by many chemists. Physicists, however, need to quantify everything, and since they could hardly ignore chemistry the discipline of physical chemistry grew up towards the end of the nineteenth century, with the purpose of combining physical rigour with chemical insight, and to act as a general intermediary between the two schools of thought.

Tribalism has always been a powerful force, and it is no less so in intellectual matters with leaders and followers eager to proclaim their distinctive beliefs. Separate development in the various sciences, far from being the exception, therefore, has usually been the rule. Loyalties to disciplines and the ideas on which they are based can be intense. Slogans can be seen on departmental notice boards: 'Physics is all there is'; 'Chemistry is life and life is chemistry'; 'Mathematics is the door and the key to the sciences'; 'The proper study of mankind is bacteria'; 'Physics envy is the curse of biology'.

There is a nice story about a biologist who went to heaven and was delighted to see many of his old academic friends walking about in the most beautiful surroundings. There were, however, no physicists, and so he asked St Peter what had become of them. 'Don't they qualify to be here?' 'Oh yes,' said St Peter smiling, and took him to a door hidden in a hillside which led eventually to another identical heaven. There he saw his physicist friends everywhere but they could not see him, or the heaven from which he had been guided. The biologist was mystified and asked St Peter to explain. 'Well, we must ensure that everyone here is completely happy, and physicists think that they are the only ones who can make it.'

Today, there are some three hundred disciplines or fields of study, and many others have fallen by the wayside. A history of their development would be encyclopaedic, but in general the emergence of the disciplines has been sporadic, and has often been based on misunderstanding.

Chemistry, for example, emerged as a science towards the end of the eighteenth century with the purpose of examining the composition and interrelationships of every known kind of matter or substance. Not surprisingly, this huge subject area split quite quickly into two broad categories: organic chemistry, dealing with the compounds found in living systems, and inorganic chemistry, which dealt with the rest. Their separate development was accelerated by the need to develop different types of experimental techniques. Inorganic chemistry was characterized by strongly linked molecules and fast reactions between chemical species, usually in an aqueous environment. For organic chemistry, molecules were usually much bigger, insoluble in water, and their chemical reactions were relatively sluggish.

The separation of the two disciplines was relatively uncomfortable in that although most organic compounds have carbon, hydrogen, oxygen, or nitrogen in their make-up, almost every chemical element can form organic as well as inorganic compounds. But much more importantly, organic chemistry faced a serious problem over which controversy raged throughout most of the nineteenth century. It concerned the nature of life, and the organic chemists' inability to synthesize any constituent part of a living system.

Their failure must have been acutely embarrassing, as a scientist's purpose is to understand Nature, and yet it would seem that although they could study the repertoire of a molecule's interactions in the laboratory, it would have no relevance to the ways molecules performed when they were playing in life's symphony. To make matters worse their lack of success reinforced the long and widely held belief that their task was impossible since life required a 'vital force' for its creation. The vitalists were a powerful group. They would seem to have much in common with their alchemical forefathers and their search for the philosopher's stone in that they believed that the behaviour of the elements inside a living organism was governed by factors outside the 'mechanistic' realms of physics or chemistry, forces which were responsible and essential for the generation of life itself.

Some help came to hand in 1828 when Frederich Wöhler discovered that urea was produced when ammonium cyanate, an inorganic compound, was heated. Before that time, urea had been found only in living organisms, and although the discovery launched organic chemistry as a worth-while discipline, it did little to convince the vitalists. Using a type of argument that will have a familiar ring to all pioneers, they said that urea was a special case since it was an end-product in the chain of living processes and was therefore not essential to life itself.

The vitalists' conversion had to wait until the end of the century when, after a long dispute, the sober truth was revealed to them, as it has been revealed to many over the ages, *in vino veritas*. Brewing has been practised throughout most of the history of mankind, but it has been understood for less than a hundred years. By 1800 or so it was known that a sugar solution could be fermented into alcohol and carbon dioxide only if it contained yeast. The vitalists claimed that the yeast must be living so that it could bring the vital force into play. Even the great Louis Pasteur subscribed to this view, but in 1897, Eduard Buchner, a German organic chemist, showed that fermentation occurs even if the yeasts are ground up with fine sand and their extract added to a sterile sugar solution. Buchner won one of the earliest Nobel Prizes (in 1907) for this beautiful experiment, and the discovery that the essential component contained in the yeast was 'zymase', from the Greek word for leaven.

The vitalists' conversion did not occur overnight, of course, but Buchner's discovery revealed yet another example of Nature's passion for universality. He showed, in effect, that molecules will be molecules whatever the environment in which they find themselves, and that their behaviour will be the same in living organisms as elsewhere. Thus, after centuries of prejudice and misunderstanding, another intellectual barrier to progress had been removed.

It was Emil Fischer, however, who showed that Buchner's zymases ('enzymes', as Fischer called them) are indeed essential to life. Organic chemistry had long been characterized by slow reaction times, but in the presence of exactly the right type of molecular enzyme, reaction rates can be increased by enormous factors. When a particular enzyme comes across one of the molecules with which it is intimately affiliated, it changes the molecule's shape in such a way that it can take part in a reaction that otherwise might be impossible or exceedingly slow. Having accomplished this catalytic task, the enzyme slips away unchanged to find a new molecule to help.

The performance of enzymes in living organisms is the subject of intense study today: their spatial structure, functions, the ways they are switched on and off, and in general mediate and control the complex chain of processes essential for the maintenance of life. Emil Fischer won the Nobel Prize for chemistry in 1902 for his wide-ranging work on the sugars—glucose, fructose, etc., their structure and stereochemistry—and on the proteins in general, of which family the enzymes are an important member. Not surprisingly, he is now regarded as the father of biochemistry for this pioneering work. Thus organic chemistry was partitioned, and the two broad components have moved progressively apart ever since.

This brief sketch outlining some of the ways in which chemistry has developed is far from unique. Similar stories can be told about every area of science, and new areas are evolving at an increasing rate. Until relatively recently, the expansion of scientific enterprise (which in this book embraces all aspects of science, engineering, and technology) and the attendant growth in its complexity has occurred largely as a result of man's innate curiosity and need to understand more of the world we live in. Throughout the past hundred years or so, however, and particularly through times of war, these largely intellectual pressures have been progressively reinforced and have often been replaced by the need to satisfy the more material requirements of industry, the military, and society in general. As we shall be discussing in later chapters, art and technology refined by trial and error have frequently foreshadowed comprehension, but there is little question that a deep understanding of the sciences on which a technology is based can substantially increase its effectiveness and reliability. And, of course, science has produced a vast richness of surprising discoveries, many of which have transformed the industrial scene.

This prolific material harvest and the high expectations of new crops to come have created pressures that have revolutionized the pursuit of science. Virtually within living memory, science has moved progressively from providing a focus of attention for a devoted and largely privileged few to being the basis for one of the world's largest and most important enterprises. Today, science in its many manifestations commands the professional allegiance of millions of people, expenditures of hundreds of billions of dollars, and has given rise to a multitude of authorities that preside over almost every aspect of a scientist's professional life.

In some respects the rapid growth of science is unremarkable: so many institutions, industry in general, commerce and finance, government, and indeed population itself, have exploded in size during the past century. Furthermore, expansion on these scales is usually accompanied by increasing levels of regulation and control, but perhaps most would agree that what we lose on the swings of personal freedom we generally gain on the roundabout of increased prosperity. Unfortunately this simple trade-off does not work for science, simply because the object of the exercise is to expand understanding at the expense of ignorance, and any restrictions on freedom can only prolong and protect the survival of that sorry state. Nobody likes rules and regulations, of course, but for natural scientists (as opposed to engineers and the like: see Chapter 4) the only ones that matter are those by which Nature organizes things; those imposed by any other authority can only cause confusion, and generally get in the way.

This problem is becoming very serious, and it is ironic that it stems entirely from success. For many years, science and technology have provided a steady stream of major discoveries that have improved, enriched, and ennobled life, at least for most people in the industrialized world, way beyond what would have been comprehensible a century ago. This glittering performance has, however, been accompanied by the growth of two factors that curtail freedom and threaten the future of scientific enterprise in general. As outlined above, scientists have coped with the problem of expansion by progressively fragmenting their enterprise into an increasing number of disciplinary domains, each enclosing a sector of intellectual territory that, for the time being at least, can reasonably be encompassed by a single individual. Inevitable though it may have been, fragmentation also tends to obscure the axiomatic assumption (see page 153) that Nature has but one dominion governed by one all-pervading set of laws. Every competent scientist is, of course, fully aware of this, but nevertheless it still leads to serious problems, as we shall see.

The second factor concerns money. Much of the enormous expansion in the academic sector has generally been funded by governments, although nowadays industry too is becoming increasingly involved. But any money governments spend is raised by taxing you and me; and being publicly accountable thereby, governments have had to devise procedures that can be shown to be even-handed. In principle, therefore, any scientist with a good enough research proposal will be funded. For much of this century that has generally been the case, although scientists have hardly ever had as much money as they need. But success has changed all that, and scientific enterprise is now so extensive and sophisticated that scientists need considerably more resources than even the richest countries can reasonably be expected to provide.

For governments, there is nothing new in this: they are all too familiar with having to budget for deficiency funding. One might have expected, therefore, a wide variety of national approaches for dealing with the growing funding famine, but science is a global enterprise, and even though governments have launched countless enquiries, debates, and reviews, have consulted funding bodies and professional agencies in profusion, and in general have left no bureaucratic stone unturned, their conclusions have hardly varied. Shorn from preamble, the outcome usually boils down to words like 'priority', 'focus', or 'concentration', which express the recognition that since no nation (or foundation, or council, or other funding body) can afford to support every line of scientific enquiry proposed by its members, we must concentrate resources on those areas that are likely to yield the most benefit.

National funding bodies have subsequently sought advice from the best and most experienced scientists, and indeed, continue to do so as they agonize over priorities and explore every possible way that quarts might be extracted from pint pots. But the global dimension is inescapable, and those areas agreed to be the best bets in one country are inevitably chosen elsewhere. As a result, the scientific agenda of most nations are broadly similar, and tend to favour such topics as biotechnology, information technology, superconductivity, electronics, and new materials, although the richer nations also try to make special provision for the big expensive sciences like astronomy, space, nuclear fusion, and high-energy physics.

Thus, in the short time of a decade or two, even-handedness has given way to discrimination. The chances of getting support for an idea nowadays will usually depend on such factors as how the work would relate to areas of national focus, or whether the work would be done at a centre of excellence or similarly favoured institution, and the prospects of a beneficial outcome. Scientists have therefore lost a great deal of their personal freedom. Even in the richest countries, almost all available resources are earmarked for fixed, long-term commitments such as national laboratories, or priority areas, or for the special initiatives set up from time to time such as mapping and sequencing the human genome, advanced materials, or a host of others, and there is very little left for research based on personal initiative.

The scientific landscape has now reached a stage in its complex evolution where exploration is being managed and controlled largely along pseudo-business lines, with boards of scientific managers presiding over their respective disciplinary divisions, competing interminably with each other in the committees and corridors of power and trying to show that investment in their particular patch will yield a better scientific return than their competitor's. These managers cannot *order* their academic colleagues to do anything, of course, but they can and do decide which of them will be allowed to do the research that their boards have agreed should have highest priority, and the feedback in this loop can be very powerful.

Science is therefore losing diversity and flexibility, and in effect its administrators are unwittingly making it easier for Nature to preserve the secrets of its vast undiscovered tracts simply because explorers cannot know in advance what benefits or otherwise their discoveries might bring.

The best scientists are painfully aware of the growing crisis, but they seem powerless to do anything about it. They know that almost every major discovery made in the past has come from what today would be called speculative research, and that very few of the scientists responsible for these discoveries would have had either the ability or indeed the inclination to provide a convincing argument that their work was going to be worth while, *before* they were allowed to do it.

Albert Einstein, writing about the motivation of great scientists, said that their long-sustained effort is not inspired by any set plan or purpose. 'Its inspiration arises from a hunger of the soul.' Even when inspiration yields dividends, a great deal of development is required to convert the new understanding into tangible benefit, which usually requires creativity and commitment comparable to that which led to the original discovery. Furthermore, its direction can often surprise the originating scientist. Some of the most important discoveries this century were made by Ernest Rutherford. Within the first decade his ingenious experiments (mostly carried out at McGill University in Montreal) had revealed that radioactivity was Nature's way of transmuting the elements, a discovery which controversially challenged the firmly established belief among contemporary chemists that the elements were immutable. He was soon forgiven, however, and Rutherford (a physicist) was awarded the Nobel Prize for Chemistry in 1908. After his move to Manchester, he made in 1911 perhaps his greatest experimental discovery when he proposed the nuclear model of the atom, in which a massive central nucleus is surrounded by 'planetary' electrons. In his Gifford Lectures, published in 1928, Sir Arthur Eddington, the British astronomer, said that Rutherford's discovery (of 1911) had introduced 'the greatest change in our idea of matter since the time of Democritus'. Indeed the word 'atom' was derived from the Greek *atomos* which means 'indivisible'. Nevertheless, Rutherford said with characteristic frankness in 1933 that 'The energy produced by the breaking of the atom is a very poor kind of thing. Anyone who expects a source of power from the transformation of these atoms is talking moonshine.' Unfortunately he died in 1937, before the harvest of his insight had been fully reaped. Only two years later, for example, the centuries-old puzzle of how the Sun generates its prodigious energy was finally solved (in principle at least) by Hans Bethe when he published details of the 'carbon cycle' by which hydrogen is circuitously transformed into helium. Rutherford would have been delighted, and perhaps would also have understood that when it comes to charting the future, policy-makers and scientists are in the same boat.

If we accept that Nature's supply of riches is not about to dry up, how best can we tap this bountiful source? For development, that is the steady improvement of an idea to knock it into better shape for use in the everyday world, this question is well understood and much industrial research is dedicated to such ends. But is it possible to control access to the sort of major discoveries that have laid the foundations of today's high technology, discoveries that led, say, to the new electronics, the laser, and genetic manipulation?

Not so long ago, this question would have been incomprehensible; how on earth could policy-makers of the 1930s and 1940s have focused on the research priorities that would lead to such revolutionary breakthroughs? In many instances even the words we use today, such as 'laser', for example, would have been meaningless to them. Most scientists today would agree that the future holds major surprises in store for us, and also that specific revolutions cannot be predicted. In these circumstances, therefore, it would seem that the best strategy is to foster diversity in research, and to allow serendipity to take its course. While few would disagree with this simple logic, it would also be considered hopelessly naïve since its implementation would require much more money and resources than are available.

Scientific enterprise, as it is presently organized, is therefore facing a depressing future. As attention is concentrated on favoured fields and disciplines, each unit of discovery, so to speak, progressively becomes more expensive as each rich vein of scientific potential becomes depleted, and ever more sophisticated and expensive equipment is required to extract the remaining nuggets before someone else does.

There are very few institutions that have survived for centuries without radical change, and it is perhaps astonishing that an enterprise which has brought about such enormous transformations in our lives has itself in at least one major respect hardly changed at all. Although its expansion has been extraordinary, present-day science has the same basic structure that it had more than two hundred years ago. The disciplines have evolved, of course, but the philosophy implicit in the early partitioning of natural philosophy is much the same as that which has underpinned every successive fragmentation ever since, a philosophy that has been dominated by such arbitrary considerations as similarities in behaviour, or techniques, and by prevailing patterns of understanding.

Over the years, the mental processes responsible for all this have become so deeply entrenched that the disciplines dominate everything we do. So much is written and spoken about disciplinary *boundaries*, about the importance of interdisciplinary (or multidisciplinary or transdisciplinary) research, because new ideas often occur between or across disciplinary *frontiers* that a casual observer might think that these barriers had real meaning or significance. Unfortunately, in at least one important respect, they do, because almost every funding agency disburses support for research and for teaching almost entirely along disciplinary lines.

Although scientists do not need to be told that the disciplines have no tangible existence, and that they are merely very convenient figments of our imaginations, there is something of the Jekyll and Hyde about the ways in which they are obliged to behave, simply because their careers would suffer if they were to do otherwise. Even for the most basic academic research, senior scientists are dragooned into attending seemingly endless rounds of meetings to advise on the fields, disciplines, special initiatives, centres of excellence, and other target areas that on balance are the most promising, and which would therefore be the most deserving recipients of the limited resources available. But none of these *entities* can possibly have promise; that prospect can only be created by *people*.

For myself, I began to recognize the potential of this deceptively simple observation only when I gave up full-time involvement in hands-on bench research. I was not entirely happy with the prospect of losing contact with my friends and ending my association with the hurly-burly and excitement of high-energy physics, but the offer, which came completely out of the blue, of becoming an assistant to the UK Government's Chief Scientific Adviser, first Sir Alan Cottrell and later Dr Robert Press, was attractive. My own tightly knit research group of fifteen would shortly be dispersing, and we were already considering membership of a small scientific army that would be more than a hundred strong. It seemed, therefore, that it might be time for a change, especially as one could hardly choose a better place than the Cabinet Office to start a new career. Three years there were followed by two years at the headquarters of the Science Research Council (now called the Science and Engineering Research Council), and following my first experience of being headhunted, two years at the Bank of England as their senior adviser on printing technology and banknote security.

My travels along the collusive corridors of power had at least one happy outcome, and that was the meeting of Jack Birks, a managing director of the British Petroleum Company, and Robert Belgrave, one of his senior colleagues. In 1979, BP had set up a committee to consider a 'Blue Skies' research initiative, and Jack Birks, recalling my growing interest in the governance of science, rang me at home one evening to ask if I might be interested. I was! At that time, I had not heard of the delightfully vague 'Blue Skies' epithet, but it seemed that BP wanted to do something 'specifically innovative' in their words, that might lead to the generation of new types of business opportunity outside their main areas of interest of oil, gas, and energy production in general.

BP accepted my recommendation on the structure of the proposed Blue Skies enterprise, and particularly that it should be completely independent of the company's mainstream research and development organization. But although I preferred the more flamboyant 'Blue Skies'—I gave my small boat that name—I had to accept that it should go under the more prosaic and somewhat ambiguous title of the 'Venture Research Unit'. Our use of the word 'venture' was in the classical sense used by Lord Byron to describe those who would have or risk a journey, 'as all must do who would greatly win'. Today, however, it is a pity that as generally used the word is almost invariably associated with capital, speculation, and gambling, and much of the spirit of adventure it had seems to have been lost.

We set up shop in April 1980, but two years later, both Jack Birks and Robert Belgrave had retired, and I was in the perilous position of having lost both my patrons. Luckily, Jack Birks was succeeded by the equally far-sighted Robert Malpas, who encouraged me to build on the foundations his predecessor had helped me to lay. Indeed, throughout almost the entire decade of the 1980s these men ensured that the Venture Research Unit had freedom to develop the initiative, and moreover, that it had all the resources it required. As far as considerations of research were concerned, we were in the astonishing position of having, in effect, *carte blanche*. In addition, to ensure that we did not jump too far from the rails, we were given an Advisory Council of glittering and stratospheric distinction. It was chaired by Sir James Menter, then Principal of Queen Mary College, London, a non-executive director of BP, and a metallurgist by trade. Its other external members were Sir Hans Kornberg, a professor of biochemistry at Cambridge, who later became Master of Christ's College; and Sir Rex Richards, Warden of Merton College, Oxford, and a professor of physical chemistry who went on to become Vice-Chancellor there. They were joined by Professor John (now Sir John) Cadogan, who was BP's research director; Robert Belgrave initially, and later by a succession of BP's senior policy advisers; and by Oscar Roith and later by his successors as BP's chief engineer.

I decided from the beginning that we should concentrate on people. But the $64 000 questions were who, and how? There is never a shortage of people who can use more money, but I had no wish to be associated with a scheme that would merely add incrementally to the pot, and to what was being done. However, after talking to hundreds of people before and after joining BP, and the usual agonizing that follows when one's bluff is called by being given the tools to do the job, we decided on a research strategy that would provide all comers with a commodity that has recently become very scarce, and which in the past has been richly productive: that is freedom, and particularly the freedom radically to challenge received wisdom. Thus, we would ignore all consideration of fields or disciplines, or boundaries of any kind, including the geographical ones, or of status, or initially of products or processes, however fantastic they might be, for otherwise the research would become mission-oriented. We would concentrate on finding those hungry souls whose heretical ideas could be brought to fruition only in such an unconstrained environment. In the past, such people have changed our perceptions, and have redrawn the limits of what is possible. If we could also integrate their work with industry, we might be able thereby to bring about the new and unexpected types of business opportunities we were supposed to create.

In research, almost every major discovery has initially been seen as heretical, and to the extent that they break from dogma, that is precisely what new discoveries are. I shall never forget my feelings of resentment when as a young physicist working at the University of Alberta in Canada I first heard news of Murray Gell-Mann's radically new theory about the so-called 'elementary' particles, the fundamental building blocks of sub-nuclear matter. In 1960, many hundreds of particles were thought to be elementary: the list included particles like the proton and the electron, of course, but also a host of mesons and baryons, each thought to be indispensable entities, and new ones were being discovered virtually every day. It was a most unsatisfactory situation in that there seemed to be no order to Nature's behaviour, and no limit to the number of particles that seemed to be elementary.

Gell-Mann, an American physicist, changed all this, and in a most elegant theory called 'The Eightfold Way' (an Israeli physicist, Yuval Ne'eman, independently and simultaneously derived a similar theory) proposed that almost every one of the current profusion of particles was composed of varying combinations of what were then hypothetical particles called quarks, of which three distinct types were proposed. Moreover, he pointed out that one other very important particle—the omega meson—had been missed, and he gave precise directions on where to find it, and predictions about its properties.

Shortly afterwards his claim was triumphantly confirmed, providing powerful evidence in favour of the new theory, which subject to modifications—the number of types of quark is now thought to be six, of which five have been identified—still stands today. However, one of the properties postulated for quarks is that they must carry one-third of a unit of electric charge, but my training over many years had deeply etched in my mind the idea that electric charge, like the atoms of ancient Greece, could not be split, and could exist only in unit quantities, including zero of course. Gell-Mann was almost completely unknown in 1962, and yet he was saying in effect that all the great men and all their textbooks and lectures were wrong in this respect. To make matters worse, the quarks were at that time merely the product of a theoretician's imagination. Even though I had no personal stake in the conventional wisdom, Gell-Mann's theory was like a slight to a close family member, which quite properly therefore should be ignored. But after the initial shock I found that it could not be ignored, because Gell-Mann's world was so much more beautiful and elegant than the one he had challenged. It was a chastening experience, and a very important step in my education.

There are very few people like Gell-Mann, who went on to win the Nobel Prize for Physics in 1969, but their impact can be enormous. Yet he is typical of those who would have considerable difficulty today in convincing funding agencies of the viability of their ideas before they become accepted. We therefore decided that our Venture Research enterprise would be dedicated to finding them, to giving them our support and encouragement, to developing a spirit of community among them, and to fostering new types of relationship with industry. The criteria we would use for their selection would be the excitement, breadth, and degree of challenge that the research offered to established order, and like all good explorers we would defer all analysis of the potential value of the new ground until it had been shown to exist.

So much for our strategy. We shall see in later chapters how we set about its implementation.

2

Choosing a problem

One of the most difficult questions for any researcher is 'What shall I do next?' If you are a beginner, it is particularly important because the problem you select for your maiden voyage will probably decide the direction of the rest of your career. In principle, the variety of choice is enormous, but the prospective explorer will quickly find that the choice is rapidly whittled down to a very small number of highly specific areas. This is because although there are thousands of researchers working at any major university or research institute, most are located within specific departments: physical chemistry, say, or molecular genetics. A typical university may have fifty or more departments, and the type of work that each can readily do will be tightly prescribed and quite strictly enforced.

Indeed, the very existence of departments is hardly compatible with the spirit of universality embedded in the university concept, for they inhibit the free movement of people and ideas. They do not *forbid* movement, however; established people from different departments may collaborate, and many do, but this is uncommon. One of Venture Research's most successful collaborations involves two departments at the same university, but although everyone directly concerned in the work agrees that a single location would be desirable, there has always been the problem of where that location would be, and which department would have jurisdiction over it. Staff transfers would be involved, and the losing department would be correspondingly weaker (and the winner stronger) in all future negotiations. There is every incentive therefore for the department that thinks it might lose to oppose the move.

Departmental wars in universities can be very unpleasant, because unlike industrial or government laboratories there are often no bosses with the powers to impose a solution. University presidents, vice-chancellors, and their attendant deans and professors rule largely by consent, and conflict can rumble on if the combatants are evenly matched in terms of resources and intellect. When times are hard, the battles can be bitter struggles for survival. One of the most bizarre

I came across was at a very prestigious university where the departments of Mathematical Biology and of Biological Mathematics were at loggerheads after a sensible and brave dean had decided that such fine distinctions were an expensive luxury.

If you were a young researcher you would hardly be aware of these high politics, and in any case the choice of department will be strongly influenced by your initial training. There will be scope for manoeuvre since, say, a physics or chemistry graduate would be allowed to choose a biological department or an engineering one. The next step might be to walk round the departmental research groups; but only a few may have vacancies, and in the end only one or two might appeal in terms of the type of work they are doing, the equipment they have, or whether you think you can get on with the people there. All this will happen quite quickly, and then the choice will be made, either by you or perhaps be forced by circumstances. So in the space of a week or two your horizons will have shortened, and you will have moved from the panoramic position of having the scientific world at your feet to one where your feet will be planted firmly on the ground in membership of a group looking at such subjects as quantum tunnelling in semiconductors, or energy-transfer mechanisms in photosynthesis. The daunting thing is that if you stay in research and make a career of it the chances are that you will be doing similar work in ten or even twenty years' time.

Although your subject may at first appear narrowly defined, you will soon find that there is a vast literature that you will need to read to bring yourself abreast of the current situation. This can be a struggle, because most scientific literature is not really meant to be read, but to record the fact that a person or a group has made an observation or a measurement, and thereby was the first to do so.

My own first experience was awful. After graduating in physics at the University of Liverpool, my departmental professor, H.W.B. Skinner, suggested I might join a group in the Nuclear Physics Research Laboratory led by Dr L.L. Green and Dr J.C. Willmott. When we met, they gave me a paper which they said was the foundation for everything they did. It was dreadful. I understood as much after the first as after the hundredth reading. I didn't even understand enough to be able to ask a sensible question about it. The words and the endless equations meant absolutely nothing to me, and I thought that perhaps I should give up research: if I didn't understand the foundations what hope would I have of building something new on them? After some time, I confided in my good friend and fellow research student Eddie

Baart, a splendid no-nonsense South African who called our mentors by their first names long before I had courage to do so. 'You're an idiot!' he said in his strong South African accent. 'That paper just has a few useful equations in it. No-one ever looks at the rest.' John Willmott mentioned some time later that he had given me the paper to take me down a peg: I was too cocky and needed to be taught a lesson. He did: I learned that knowledge should never be used as a weapon.

Whether your first problem has been chosen by yourself, or allocated to you, it will quickly become your own as you become immersed in it and first realize that you have a real chance of doing something or thinking a thought that has occurred to no-one before. Thus, the problem not only becomes yours, but you are associated with it, first within your own small circle, and then wider afield, as your reputation gradually grows.

My own problem was to understand the nucleus of the sodium atom; not just any sodium nucleus but the ones that have 11 protons and 12 neutrons: sodium-23. Eddie Baart was looking at two nuclei: phosphorus-30 and aluminium-26. Both these nuclei are unstable, but more importantly, the fact that they have an even number of nuclear constituents—nucleons—meant that their behaviour would not have the same character as my own sodium-23. This is because for an even nucleus pairing effects would be expected to be different from those of an odd one, where there will always be a 'three's company' situation somewhere that leaves a lonely wallflower on the nuclear dance floor. Singles behave differently from couples, and if you wished to study their social behaviour you would need to take care that your own intervention did not introduce a distorting influence. Thus, if you are dancing with a partner in the old-fashioned way, and catch the eye of one of these single wallflowers you become a new couple, perhaps only for a fleeting moment, but during that time your partner becomes the wallflower, and the behaviour of all concerned will be affected. The effects may also depend on whether you are a man or a woman. Nuclear physics is no less subtle.

Although we were aware of each other's problems, and those of our friends in other groups, it was surprising how quickly one's own research project becomes all-consuming, and one's personal universe revolves around it. When you become a scientist, therefore, you will find that, in the space of a few years, you will have become an expert, and people will seek you out to ask your opinion. Although it will have been a hard and often lonely slog, you will find that it will have been worth while, especially for the inner satisfaction of being perhaps the only person who really understands some small aspect of the great big picture.

Following your ordeal by Ph.D. examination (see Chapter 5) you will again have to decide where you want to go: to stay on if you have the option, or to go to another university, or to one of the many government or private institutes where research is done, but which do not usually have teaching responsibilities. This time your choice will be strongly influenced by who you know, or more likely, who your supervisor knows, and who has the money to take you on as a 'post-doc', as post-doctoral researchers are usually called. Up until this time you will have been paid only a subsistence allowance, but now you will be paid a salary which may be two, three, or four times what you have been getting, depending on where you have come from and where you go. As will be discussed in Chapter 5, you should not expect anything permanent at this stage, and you will probably have to move every two or three years until you find a niche for yourself. The research you will do during this itinerant phase will largely be determined by the interests of those who have the money to pay you. They in turn will probably be looking for expertise applicable to one of the current mainstreams because that is where most of the money is. You are now the proud possessor of a track record, and research leaders will feel more comfortable trusting you to do something if you have done a similar job successfully before.

Non-scientists will also be familiar with this idea; it is only the most exceptional employer who will appoint someone claiming, however plausibly, to be able to rise to a new challenge in an area in which they have no experience. But for the searcher-after-truth, building up a track record can be a tender trap, and increasingly you will find that what you have done in the past will tend to dictate what you will be allowed to do in the future. Unfortunately, the truth is hardly ever where you expect to find it.

The achievement of specialist status in one or a few of the many disciplines of science is nevertheless usually regarded as an attractive prospect rather than a trap, just as it is in many professions or guilds such as the law, finance, or plumbing. Once achieved, it is often the case that the grass is browner on the other side of the fence, and one's own calling is generally seen as more interesting and important than the others. Nature, of course, is above all this, and will not adjust its behaviour to suit the rules of the club that claims your allegiance. The best scientists are well aware of this problem, and very soon learn to correct for the distortions introduced by their specialist training.

I am, however, getting ahead of my story. It has been implicit in what has been outlined so far that our main concern has been with Nature, and our attempts to understand the world of which we are all a part. Since we *are* a part of it, we should always be aware of the distortions that might be introduced by our own individual senses and perceptions. If the object of our attention is as impersonal as an atomic system or a galaxy, one might expect these aberrations to be at their lowest, but even here, as we have seen, prejudice can be powerful, particularly when it is seared into our subconscious, and everyone around us seems to have a similar belief. Nevertheless, although Nature is complex, she is not devious, and sooner or later someone will notice the flaws in our philosophy and will be able to set up an experiment—which is in effect a dialogue with Nature—to show that the radical view is more accurate than the one that had been accepted.

If the object of our attention is a human brain, we can see that this simple picture may get more complicated. If we study the brain's involuntary functions—the control of our breathing, heart, digestion, or other metabolic processes—we should expect to get much the same results as we might from studying the brain of any other animal. If our study involves asking a question, we know that, if it is clear and unambiguous, Nature will always give the same reply in the same circumstances. If the question involves human voluntary functions, however, the concept of purpose is raised, and answers might vary among individuals, or with time. A well-known story involves the question 'What is 2 + 2?' A mathematician replies, '4'; a scientist, '4.0'; an engineer, 'between 3 and 5, but let's call it 6 to be on the safe side'; while an accountant might answer, 'What number did you have in mind, sir?'

Scientists should therefore expect to find new complications in addition to those found in Nature whenever humans play a purposeful role. Therefore, of all the current divisions and departments, fields, and fashions found in scientific enterprise the ones that separate the so-called natural sciences from the sciences of the artificial—as Herbert A. Simon has called them—seem to be different from the rest. Simon is Professor of Computer Science and Psychology at Carnegie Mellon University in Pennsylvania, and won the Nobel Prize in Economics in 1978. The 'sciences of the artificial' include engineering, computation, economics, and indeed any science that describes the ways in which complex information is processed by people. Not surprisingly, these definitions tend to be controversial. Some economists, for example, strenuously deny that they are scientists since, among other things, they would claim that it is difficult or impossible to carry out controlled experiments. If a government wishes to know how, say, an increase or decrease in an exchange rate might affect inflation they would not expect to be asked to await the results of an experiment that tested that particular water. Economists would thus say that they are observers of complex systems, and develop theoretical views about how people and markets should behave. Astronomers might in fact make similar comments, but I have not yet met one who would deny being a scientist.

These subtle differences in the sciences affect the ways in which research is carried out, and the problems that are chosen. It seems sensible, therefore, to consider them separately, as will be done in the following two chapters. Despite the subtleties of these distinctions, hungry souls will nevertheless find considerable nourishment in either of these sciences, or both.

3

Playing with Nature

In 1985, Harvey Scher, who worked at the Warrensville laboratories of what is now BP America, phoned to invite me to speak at an American Institute of Physics meeting in New Jersey. The occasion was the annual meeting of the Corporate Affiliates, a programme that brings together senior industrialists and academics to exchange views on the potential of R and D, and Harvey Scher was always on the lookout for opportunities that would allow the glories of Venture Research to be expounded. The meeting was held at the Corporate Research Laboratories of the Exxon Corporation, which location had been agreed some months before. Unfortunately its timing turned out to be embarrassing as only a few days before the meeting Exxon had announced a large number of redundancies there; during the lab-tour the atmosphere was gloomy, and the empty benches were conspicuous. Exxon did us proud, however, and the facilities were magnificent. The talks were well attended and offered a wonderful opportunity, therefore, not only to help extend Venture Research interests in North America, but also perhaps to find one or two other companies who might wish to collaborate.

After my talk, a chap from the audience whom I had noticed was taking lots of notes, came up to me with great enthusiasm. 'I can hardly believe what you have said. I feel as if I've died and gone to heaven.' He was Dudley Herschbach, whom I had not met, but I knew that only a few days before it had been announced that he had won the Nobel Prize for Chemistry. 'If you've got the time, I'd like to talk to you about a great idea. No-one will take it seriously. I'd swop my Nobel Prize if I could get support for it!'

What an opening gambit! I was, needless to say, taken aback. I had met Nobel Laureates and other VIPs before, some of whom had behaved as if I should be honoured that they were willing to take money from my organization. Invariably, money requested in this way is for some expensive piece of equipment, and it takes considerable time and energy to convince them that Venture Research is much more than a supply of money for people to do more of what they have done before. They must want to break out in some way.

We have always taken a pride in Venture Research in taking our decisions on who to support exclusively on the quality and potential of the idea—whoever it comes from. My concern was, therefore, how could I respond to this chap in the same way that I would to a post-doc? He positively radiated brilliance and patient good humour, and, had he been an ordinary mortal he should have been at the height of his Nobel euphoria.

My concern slowly evaporated as I got to know him. His passions— science, which for him is natural science, and people—are indistinguishable. Both provide him with endless opportunities for amusement and stimulating conversation, but for science there is a fascinating complication in that, as he says, 'Nature speaks to us patiently in many tongues, most of which we understand only imperfectly.' Nature is not malicious, but makes no concessions to our often considerable lack of linguistical abilities. Enquirers may be given the same message over and over again, but unless we are smart enough we may mishear, misunderstand, or even fail to recognize that there *is* a message.

After he had won the Nobel Prize he had received hundreds of letters of congratulation from friends all over the world. He says he has little skill in human languages, but he could nevertheless work out the gist of what they said—except one, which had him stumped. 'Can you guess', he said, beaming broadly, 'the language that it was written in?' The anticipated suggestions of such languages as Chinese or Arabic were greeted with a bigger smile, until 'Braille' was proclaimed triumphantly. 'Now, how do we know Nature is not speaking to us in her equivalent of Braille, or something even more esoteric?'

It is an excellent question. Almost invariably science students are trained to converse in one or perhaps two of the relatively old languages of physics, chemistry, biology, mathematics, etc., and some of their dialects. Multilingual abilities are not encouraged, and most scientists behave more and more as if Nature speaks, so to speak, in only one or two of them.

One of the questions I like to ask of prospective Venture Researchers is 'What would a Martian scientist make of what you have just described?' Martian students are trained, as the imaginative know well, in quite different types of schools and colleges; they think differently, and their senses too are not the same as ours: they can 'see' a magnetic field, for example. Nature, whatever they may call her, is the same for them as for us, but their scientists work in different disciplines, and give much higher priority to living systems than we do. Consequently, their science and technology has a distinctly Martian grammar, as opposed to our Earthly one. Being from a red planet, they can see only in the infra-red. They cannot see blue at all, and so they may not be aware of the stars. Astronomy played an important role in the development of terrestrial science, so it would be interesting to know what it was that first caused Martian shepherds to wonder. One can go on in this light-hearted vein, but who believes that our own senses are perfect? We all have our blind spots, physically and intellectually, and who knows what they close our minds to?

Herschbach had won his Nobel Prize for his pioneering work on the kinetics of chemical reactions. Thus, he was more interested in *how* chemical reactions proceed than in their end-products. In the 1960s the precise routes taken for even the simplest chemical reactions were largely unknown. This is hardly surprising, bearing in mind the economic pressures on chemists, and the complexity and speed of chemical reactions. It is often sufficient to know that if one puts certain materials into a black box other more valuable materials come out. Politicians, for example, with responsibility for health care, might see hospitals as black boxes into which streams of patients and doctors enter, and from which streams of healthy people and doctors emerge, one would hope, in the same numbers that went in. The flow might be steady, on average, from day to day, but of course the time taken for any one patient to pass through the system might be an hour, a day, a week, or months, according to the severity of his or her condition. Many politicians might be concerned only with maximizing through-put, and would take little or no interest in such details, but their performance might be transformed if they understood the broad reasons for the large variations in patient transit times, and perhaps knew something of medicine.

But we must return to a more conventional interpretation of chemical reactions. According to our present understanding (a phrase that should precede all statements in science, and in every other subject for that matter), all chemical molecules are complex. The simplest consists of two atoms of hydrogen in close association; that is two nuclei and two electrons. All electrons in all molecules are identical, and so it is meaningless to discuss which electron belongs to which nucleus, or to construct a theory based on specific identification. This subtle form of complexity pervades chemistry, and indeed all the sciences, and hints at the problems that will be met if one tries to explain molecular or nuclear behaviour using everyday experience. When two typical molecules interact, one is observing reactions between many tens or even hundreds of electrons for simple reactions, and many thousands or even hundreds of thousands for the most complex. Typically, a molecular reaction takes place in a series of distinct stages over periods of, say, picoseconds* or less. It would be impracticable at present to resolve these stages from bulk reactions in a test-tube, although some information about certain types of reactions can be gleaned from spectroscopic studies.

A picosecond is impossibly short on everyday time-scales, but to an anthropomorphized electron it might seem like an age since it could traverse more than a million atoms in that time. In principle, therefore, we might expect the problem to be soluble if a sensitive technique could be found.

Herschbach's solution was to aim beams of simple molecules at each other, and to analyse the speeds and directions of the interaction products as they shoot out from the collision zone. Although molecules can have a few relatively free electrons dancing in attendance, most are imprisoned in the vice-like grip of their massive nuclei to form a strong but flexible structure that can rotate or vibrate when it is hit. The more weakly bound electrons at the molecular fringe are relatively free to flirt with similarly flighty electrons of any other molecule they may meet. The colliding molecular beam technique can yield enormous amounts of information on the relative importance of the various electronic and molecular interactions, and on their energies and transition times. The technique has been very influential in laying down the essential groundwork for understanding molecular behaviour, which may one day make it possible to predict ways of optimizing important chemical reactions and point the way to new classes of reaction as yet undiscovered. The award of the Nobel Prize for this pioneering and seminal work, which Herschbach shared with John Polanyi (also a Venture Researcher) and Yuan Lee, who had made similar but complementary contributions, was therefore richly deserved.

Now, there is hardly a scientist who starts out on his or her career without entertaining hopes—secretly or openly—of making a major discovery or perhaps of winning a major prize. It is important that the rules are well understood, therefore, so that one may plan one's campaign. Richard Feynman, the irrepressible, extrovert American physicist was once asked how a person should go about winning the Nobel Prize. As I recall, he replied, 'It's easy, just find out what your

---

* A picosecond is 10^{-12} second or one million millionth of a second.

colleagues are doing, and then go in the opposite direction . . . But above all, enjoy what you do.'

Excellent advice, but perhaps a little more guidance is required. First, the Nobel Prize is awarded only in the sciences of physics, chemistry, medicine (or physiology), and economics, and so you will need to avoid mathematics, engineering, computing, and quite a few other subjects, or go in for literature or peace. Second, although the final selection for physics and chemistry is made by the Royal Swedish Academy of Sciences, and for physiology or medicine by the Royal Caroline Medico-Chirurgical Institute, also in Sweden, it is based on the opinions of hundreds of established scientists world-wide. You will have to publish and speak extensively, therefore, if you are to be seriously considered.

The Nobel Prize tends to be awarded for two broad types of contribution: major and unpredicted discoveries that come out of the blue (though major and predicted discoveries sometimes also qualify) and sustained programmes of innovative work extending over many years—like Herschbach's—which have brought about important changes in our understanding. As ever, luck and other factors will play their part, for there are many scientists whose work might be fairly claimed to qualify who have never won the Prize. Einstein, for example, is most remembered by the general public for his work on relativity, but if that had been all he had done he might never have been honoured. He won the Prize in 1921 for his work on the photo-electric effect, a discovery which laid the foundations for the enormous advances in electronics which have so strongly influenced us all throughout this century.

As Herschbach had barely had time to set down his celebratory glass of champagne when we first met, I could not avoid asking myself why he was talking to me? Surely he now had everything he wanted? His story turned out to be fascinating. As a professor of chemistry at Harvard, he was always on the lookout for new ways of presenting ideas to his students, whom, like many other good teachers, he sees as sources of stimulation and rejuvenation. In 1981 he had read an article in the journal *Physics Today* written in the previous year by Ed Witten, a physicist from Princeton University, on quantum chromodynamics, the intriguing name given to the theory of sub-nuclear interactions. For many years, the proton was thought to be one of Nature's primary building blocks, the end of a line of complexity, so to speak. But we should have known better in that Nature's complexity seems to be 'infinite in all directions' in the far-sighted words of Emil Wiechart published in 1896 but resuscitated recently by Freeman Dyson, a professor of physics at the Institute of Advanced Study also at Princeton, who used it as a title for a book. Protons are now known to be composite particles made up of even more fundamental particles, which have been named quarks and gluons, which are also suspected of being composite, as we should expect. However, quarks and gluons are very strongly interacting entities, composite or not, and Witten was proposing a novel route for the rough estimation of their energies of interaction based on the assumption that the dimensions of the space they occupy can be treated as variable.

Herschbach's reasoning was that if this strange technique was good enough to give even crude numerical estimates for sub-nuclear matter, where the immense forces are largely unknown, it might work quite well for calculating the energies of interaction between electrons, whose lifestyle is well known and relatively celibate in comparison with the furiously passionate quarks. At that time, he thought that the technique could be used as an exercise for his students to calculate the energy of the hydrogen molecule. As a good teacher should, he checked the calculation himself on his pocket calculator. It worked very well.

Now, it should be explained at this point that the calculation of molecular energies is notoriously difficult. Many hundreds of scientists world-wide have battled for many years using progressively more powerful computers, but relatively accurate results have been achieved only for the simplest systems. Thus we find, as in everyday life, that there are many problems in science that are easy to describe, but apparently impossible to solve.

This situation arises because the world we live in is essentially complex: many of the events we observe—the weather, the outcome of a football match, the causes of instability in a country—rarely have a simple, single cause. Rather, they are the outcome of many factors, each of which may be of comparable importance. Under some conditions, the factors seem to conspire to produce a chaotic outcome with no stability or predictable order, while for other perhaps less extreme conditions, the apparent conspiracy is not complete, order prevails, and a specific result ensues. Much of molecular behaviour falls into the latter category, as our very existence confirms every day. But the innumerable factors that would contribute to the calculation of energies using conventional means result in a complexity that is much like trying to find a sailing boat's position accurately after a long ocean voyage by keeping track of every change in speed, wind, and tide during the trip using the boat's instruments alone, with no opportunity of establishing position using the Sun, or stars, or satellites.

For a molecule like nitrogen, which consists of two atoms, there are 14 electrons, and in principle each of them interacts with the two nuclei and with every other electron. The electrons are identical, but they are not completely free. Like people in a large organization, each electron has its place in the atomic hierarchy according to its proximity to the centre of power, and both the number of positions available and the number of electrons allowed to occupy them are strictly controlled. The uncovering of the rules that govern this power game is one of the triumphs of quantum mechanics, and armed with what we now call the Pauli Exclusion Principle, it would seem that Nature rigidly enforces the terms of each electron's tenancy, and its location in the hierarchy. Exceptions are not allowed. These rules determine the limits of chemistry in every industrial plant or living organism on this planet or any other.

The challenge in estimating the total energy of a molecule is to compute all the interactions between every constituent electron and nucleus in turn, as allowed by the rules as we know them, in such ways that the energy sum of all the possible interactions is the same as the experimentally observed result. These calculations are horribly complicated. They hardly ever give the right answer, but much more important, they do not offer insight into molecular behaviour. The problem is that molecular interactions are dominated by the fringe electrons, which are bound only weakly to the nucleus. In the case of nitrogen, the bond is relatively strong in chemical terms in that it is a triple bond involving three pairs of electrons. However, even in this case, the energy to break the bond is only 0.04 per cent of the total electronic energy of the molecule, and if one wishes to calculate this dissociation energy to an accuracy of, say, 1 per cent, the total energy must be accurate to 0.0004 per cent. It is much like trying to weigh the captain of an ocean liner by taking the difference between the weight of his ship with him aboard, and with him ashore. The surprising thing is that so many people are eager to try.

It is surprising because everyone knows or should know that Nature is very subtle and does not yield to confrontation. Yet there are so many fields where large amounts of money and resources are thrown at problems in the hope that Nature will crumble before the onslaught. Dudley Herschbach's idea was certainly subtle. He had found that if he calculated the molecular energy, first on the (sublime) assumption that the molecule occupies an infinite number of dimensions, and second on the (ridiculous) assumption that molecules have only one dimension, and then interpolated between the two extremes to find the answer for molecules that occupy three dimensions, which after all is the object of the exercise, he could improve the accuracy of Witten's technique by nearly a million-fold.

This astonishing and intriguing result meant that his simple hand-held calculator could be used to yield about the same accuracy for a two-electron atom as the latest mainframe computers using conventional techniques. What on earth was Nature trying to tell him? Why is it important, apparently, to look at molecules through eyes that severely distort the fabric of space, and take us into realms we can hardly imagine? Herschbach wanted to explore this weird territory wearing these strange dimensional spectacles, but no funding agency would give him the money to do the job properly. In effect, they were saying stick to your knitting (kinetics), particularly as there could be no guarantee that his strange quest would yield anything new about chemistry.

Our attitude in Venture Research was precisely the opposite. Here was an outstanding scientist who had come across an idea that could possibly transform the way we think about molecules, and space itself, for that matter. Nature did seem to be trying to tell him something important, but in a language completely unknown to us. The idea that dimensions can be manipulated is not completely alien. The shapes of clouds, trees, or mountain ranges are complex but each of them can be described by what mathematicians call their 'fractal dimensions'. These descriptions are entirely mathematical, and we do not know whether there are other principles—physical or biological, say—that determine them. It is, however, always exciting to look at old problems from a new perspective—to take a Martian's-eye view, so to speak— and so it was not surprising that Herschbach wanted to devote the major part of his energies to it, and was so excited about its prospects when we first met.

I must admit it took some time for the penny to drop, but the idea became progressively more intriguing as the months passed, which passage of time also enabled us to test his own commitment to the idea. Eventually, after a few months all scepticism vanished, and intellectually speaking we became Herschbach's collaborators, as must be the case for every Venture Researcher. Thus reinforced, I was able to persuade BP to let us have the money we needed, and the research has since gone remarkably well.

Ernest (later Lord) Rutherford once said that when money is short there is no alternative but to think. The Venture Research concept is the result of our thoughts and those of the participating scientists. The research involved is not a next-step: it is not a logical progression from what has gone before; indeed it is a point of departure. I have described the dawning of Herschbach's venture in some detail, not because it is especially important, but because of the many illustrative points it contains. I will now conclude this chapter with a few more examples. I should, however, emphasize here a comment made in the Preface. At the time of writing, each Venture Research programme was still in a mercurial state. Rather than outlining some of the early results, and perhaps running the risk of picking the ones that subsequently might turn out to have been unimportant (or wrong!), I have limited myself to describing the perspectives that the scientists themselves would have as they opened their new doors and the main features of the landscapes they would see lying before them. The story of their adventures as they will eventually unfold must, of course, wait until a later date.

Finding prospective Venture Researchers is never easy, and sometimes months of constant searching may pass before one feels the tingle of excitement on hearing something very unusual. The next episode in this short series began when one of my BP colleagues, Dr Ken Bourne, had mentioned Venture Research to Dr Peter Day, the Director of the Agricultural and Food Research Council's Plant Breeding Institute in Cambridge, who had subsequently invited me to give a talk and meet some of his colleagues. We had a long discussion before lunch about the state of science and the future of his laboratory in particular, which was about to be taken out of the public sector and privatized. I was to give my talk at 3 o'clock, but there was time to meet Dr Mike Bennett and Dr Pat Heslop-Harrison. Half an hour later, I became their intellectual collaborator, and was determined to get them the support they needed for their work, which a few months later we were able to do.

They are interested in Nature's arrangements by which living systems control themselves and reproduce. Every large organization has its own methods for filing, accessing, and updating the information essential to its continued existence. They are rarely satisfactory, and every few years it seems that someone will try to reorganize them to cope with some new situation.

Nature's arrangements have evolved over many millions of years, and hence are more stable, even though evolution continues. Although the details vary between species, the general rules are much the same for them all. Nature writes its instructions in a four-'letter' alphabet in which each letter is a specific molecule called a nucleotide, and the letters spelling out a code are attached in the correct order to a molecular backbone, thus making up a strand of DNA (deoxyribonucleic acid). Like a blank sheet of paper, the backbone contains no information, and while the letters are different from each other they are also closely associated in pairs. The nucleotides have easily forgotten names (adenosine and thymine; and guanine and cytosine) and so for our conceptual purposes I would like to describe them in more familiar terms; one pair of nucleotides might be thought of as say bread and butter, while the other might be needle and cotton. Imagine then that Nature wished to construct a segment of DNA consisting of a five-letter 'word' spelled say:

<blockquote>bread needle needle cotton butter.</blockquote>

Nature's words are usually vastly longer, but whatever the length, there is always the possibility of introducing an error when a word is copied during replication. To safeguard against this chance Nature writes each word twice, but the message is not merely duplicated, it is written using the close partner letters. Thus, in this case it would be spelled:

<blockquote>butter cotton cotton needle bread.</blockquote>

The two words are then zipped together, and when the cell divides and a copy of the DNA has to be provided for each half (replication) the zip unfastens, and each nucleotide attracts its close associate from the cell's supply of nutrients. The two new zips can then close, and the two DNA molecules can go their separate ways. For geometrical reasons, it turns out that this can be done only if the double strands are entwined in a helix. The unravelling of this exquisitely elegant system was achieved by Francis Crick and James Watson working at Cambridge University in 1953. They shared the Nobel Prize for medicine in 1962 with Maurice Wilkins from King's College London, whose pioneering X-ray images of DNA also made a vital contribution.

Some strands of DNA contain the instructions to a cell on how to make things—a protein, or the components of a liver cell, or whatever; much longer segments containing many instructions are arranged in genes that may control a trait: pointed leaves or dark hair, for example; genes are arranged in chromosomes, some of which may control general characteristics such as sex; and a complete set of chromosomes is called the genome, which is usually regarded as the blueprint for a living species. This very simple picture has become clear only during the past fifty years or so, and although there has been considerable progress during that time, our understanding of living systems is still in its infancy, as a visit to your medical practitioner may confirm.

The unit of almost all living things is the cell. Viruses are the main exception. Sir Hans Kornberg once described them as 'bad news wrapped in protein'. Although they can replicate, they cannot do so on their own: they need a well-nourished host cell. Like well-dressed thieves at a party, they steal what they want and hope to go unnoticed.

A cell is like a factory, and it usually has a specialized function depending on its location within the animal or plant. Generally, it has a head office, the nucleus, where the genome is located, and in principle, most types of cell have the information to construct a complete animal or plant, although it is rarely activated in mature animals.

Bennett and Heslop-Harrison are interested in the nucleus, and particularly in the genome's spatial structure. As their story unfolded, I was amazed to learn that before Mike Bennett began this work in the early 1980s, it was generally believed that either the nucleus had no spatial structure, or if it did, it was not significant. This belief had apparently arisen from decades of using microscopes to study cells by sandwiching living tissue between thin glass plates. The resultant squash not surprisingly obliterates any three-dimensional structure. As Bennett explained, one might as well try to study the structure of eggs after throwing them hard against a barn door. Thus, it was earlier generally believed that the precise location of each chromosome within the nucleus was not important, and nuclear behaviour would not be affected if the nuclear soup was shaken or stirred.

I was amazed because this image is not consistent with our understanding. If Nature abhors anything it is surely anonymous uniformity. There is characteristic structure all around us: in the rocks, as we can readily see; in the oceans through variations in currents, salinity, temperature, etc.; in the air through variations in pressure, or water content, etc.; and in every living thing. Why should Nature suddenly abandon all this richness at the nuclear membrane, and develop a taste for gruel? My amazement rapidly gave way to excitement as I realized that Bennett and Heslop-Harrison were about to enter a vast, unexplored field which would give us a new perspective on life. They had already made some preliminary forays, but even though they had Peter Day's enthusiastic support, he could not authorize a full-scale attack because the work would not lie within one of the Laboratory's priority fields. Their preliminary work had been done in the evenings, at weekends, during any time they could scrounge from other programmes: 'behind the fume cupboard' in the laboratory jargon (also referred to as skunk work or boot-legging in North America). Worse still, Heslop-Harrison's appointment was about to come to an end, and Peter Day did not have money to extend it. Heslop-Harrison thought he would go to the USA, which would be good news for the Americans, but would break up a blossoming collaboration.

Their technique is beguilingly simple, as is often the case when Rutherford's advice is followed. Living cells are frozen in epoxy resin and sliced with a diamond knife with such fineness that the nucleus is cut into many sections. Each section is analysed with an electron microscope, and the data are collated and reconstructed to make up a three-dimensional picture of the nucleus. Bennett and Heslop-Harrison used a hybrid plant as their model system, and chose barley and wild rye because their nuclei have the same number of chromosomes (seven) from each parent. Although the chromosomes from each plant are comparable in size, they are sufficiently different to be distinguishable from each other in the hybrid.

Data from one of their earlier experiments can be seen in Fig. 1. What might be seen as black sausages are the considerably magnified (× 50 000) chromosomes of wild rye which surround the barley chromosomes like a shield. Furthermore, they found that the rye chromosomes are dominant, so that the hybrid tends to have rye-like characteristics; the relative position or address of each chromosome is

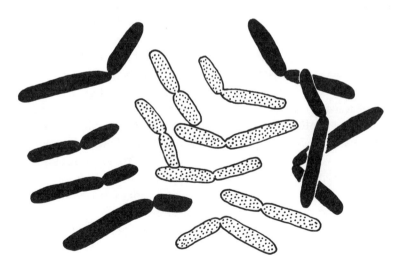

Fig. 1. A model constructed from electron microscope data of a first-generation hybrid of barley and wild rye viewed from above. The model shows the chromosomes of wild rye (black) and barley (stippled), and the tendency for the barley chromosomes to occupy a central position.

maintained throughout the life of the cell; and, as in a factory, location and function seem to be closely related. They had already begun to understand why some plant species hydbridize while others do not: if one parent's chromosome is too large, for example, it may not allow a solution to the three-dimensional topological puzzle that would make up the nucleus of the would-be offspring.

Genetic engineering techniques have been used for more than a decade, but it is still not generally understood why some genetic changes are expressed, or acted upon, while others are not. Genetic engineering is something like an application for planning permission to change the specification of a building: the local authority, that is Nature in this case, does not always approve, often for mysterious reasons. However, their preliminary results seemed to suggest that expression is influenced by a gene's genetic neighbourhood; some might be difficult and others might be easy.

The prospects for research seemed to be as remarkable as the fact that they could not get the work funded, and I was delighted to be part of it. The main problem hitherto seemed to be that they wanted to take a coherent approach to the study of nuclear architecture in living systems generally—animals and plants. However, their laboratory was authorized to study only grasses; other plants must be studied elsewhere, a consequence of the quest for efficiency and economies of scale. Studies of man would require the approval of the Medical Research Council and take them into another realm of complexity. Armed with Venture Research support, however, they could do all these things free from bureaucratic restrictions, as indeed Nature is herself.

Turning to another part of the Natural forest: like true love, the course of research never did run smoothly, and in some cases the gestation period for the germ of an idea for a substantial research programme can be as long as a decade or more. Thus, although one's attention may have been attracted at some magical moment by a curious observation, many years might elapse before one comes across enough pieces of the intellectual jigsaw to be convinced that the picture that seems to be unfolding is indeed remarkable. To be a scientist means, however, that one must be constantly vigilant, for that inspirational instant can present itself any time or anywhere, but Nature is subtle, and there will not be a fanfare of trumpets. Charles Wilson, for example, on a walking holiday in Scotland, in 1894, noticed from the summit of Ben Nevis a brilliant interplay between clouds and rays of sunlight that inspired him to produce an instrument for charged particle detection called the cloud chamber. The device led to many momentous discoveries in nuclear physics in the 1920s and 1930s, and earned him a share in the Nobel Prize for Physics in 1927.

I have long thought, however, that the prize for the person who has drawn attention to the significance of what at first glance might seem to be the most mundane of everyday observations should go to Heinrich Wilhelm Matthaus Olbers, the German physician and amateur astronomer. In 1823, he not only noticed that the sky was dark at night, but also had the amazing audacity to ask himself why that should be so. Olbers was perplexed because he knew that the Sun is merely a local star. Thus, as he wrote,

> if there really are suns throughout the whole of infinite space, and if they are placed at equal distances from one another or grouped into systems like that of the Milky Way, *their number must be infinite, and the whole vault of heaven must appear as bright as the Sun*; for every line we can imagine drawn from our eyes would necessarily lead to some fixed star, and therefore starlight, which is the same as sunlight, would reach us from every point of the sky. [My italics]

But the sky *is* dark at night, and so what might be the fallacy in Olbers's apparently impeccable reasoning? Olbers concluded that the answer lay in the fact that space was not completely transparent. If there were a multitude of stars, it would mean that, like people in a crowd, one star could eclipse many others, and if we cannot see them, their light cannot reach us. Hence the sky would be dark.

Olbers was quite wrong, in fact, and so the paradox that now goes under his name remained. It is surprising that someone clever enough to notice the darkness of the night sky did not also realize that people are unlike stars in that they do not *radiate*, at least not in the visible spectrum, and as much light will be scattered *into* our line of vision by some stars, as will be obscured by others. (Strictly speaking, the light would not be scattered but absorbed and re-emitted.) The answer to Olbers's paradox was not revealed in his lifetime, and requires Einstein's input as well as that of many others who have added to our understanding of cosmology. The answer is also very complicated, but among other things, our universe is not infinite, and the light from distant stars received on Earth is diminished by relativistic effects as well as by distance. Thus darkness at night can be explained, and it can now be seen as a dramatic indication that there were more things in heaven and earth than were dreamt of in early nineteenth-century philosophy. So, where else might the commonplace be cloaking complexity?

Anyone who has peeled a potato will have noticed occasionally that some parts might have rotted, and therefore have to be removed. These discoloured bits are collections of infected potato cells, and when Alan Paton from the University of Aberdeen Bacteriology Department put them under a microscope in 1974, he saw masses of dead cells, as countless others had done before, but he also noticed that one cell in a hundred or so was packed almost to bursting with bacteria. How can these bacteria get into a cell, he asked himself, if they cannot get out? Plant cells are normally surrounded by walls made up of close-packed molecular soldiers whose duty it is to keep out intruders, and in general to separate the inside of a cell from the dangerous outside. Paton then remembered that fifteen years before, while having coffee in the Department's lounge, one of his colleagues had brought in a pile of unwanted books. Among them Paton found a private publication from Sweden about so-called L-form bacteria and plant disease. L-forms had first been observed in 1935 by scientists working at the Lister Institute in London, who used the name L-form to describe bacteria that have either permanently or temporarily lost their ability to synthesize their cell walls. Deprived of their external protection, these bacteria are highly vulnerable, and normally would not survive. Indeed, the antibiotics—discovered by the combined efforts of Alexander Fleming, and Howard Florey and Ernst Chain in the 1930s and 1940s—are effective precisely because they destroy a bacterium's armour-plating mechanism, and so leave it virtually defenceless.

A bacterium's armour not only protects it, but also determines its outward shape; a bacterium clad only with the naked skin of its membrane can take up virtually any shape if it can survive. Recalling his coffee-lounge find of long ago, Paton thought that perhaps these bare bacteria could somehow slither their way between the cellular soldiers of a plant cell wall and take up residence inside. When they later grew, and returned to their armour plated state, as it seems they do when a cell dies, they would of course not be able to get out. But why had these awful aliens been allowed to get in? The bacteria in question are pathogens, that is they are a plant's mortal enemies, whose sole purpose is to kill. But the bacteria had grown inside the cell, which presumably had remained viable despite the presence of its lethal guests. Had he stumbled across a new type of plant–bacterial symbiosis or mutually beneficial relationship? If so, what was it?

The questions came in floods, but unfortunately the means that would allow him to begin finding answers did not. The funding agencies could no more believe that bacterial pathogens and plants could enter into cosy compacts than that lambs could lie down languidly with lions. Furthermore, plants and bacteria belong to fundamentally different categories of organisms, and conventional wisdom would regard them as closely related as chalk and cheese. So, if Paton wanted to satisfy his curiosity, he had little alternative but to manage with only the very limited resources available in the departmental laboratory. Even so, he managed to collect what seemed to be several pieces of a new jigsaw that indicated to him that L-form bacteria *might* be entering into close relationships with the plants, and that the resultant new plant–bacterium species, neither wholly plant nor wholly bacterium, *seemed* to be resistant to the 'fully metal-jacketed' bacteria normally found in the wild.

Luckily, Alan Paton heard about Venture Research from one of his university colleagues, and I met him in 1983 to hear about these remarkable findings at first hand. It was not easy to make up our minds, not because what he was saying was difficult to understand, but because his work had been plagued by uncertainty and there were problems with the reproducibility of results. Looking back, I can see that, for perhaps the first time, I was beginning to understand some of the enormous difficulties of setting out to test the validity of a specific conjecture in biology. In contrast with research in, say, physics or chemistry, where it is usually possible to isolate the system one wishes to study, living systems like plants, bacteria, or you and me, are not only breathtakingly complicated, but attempts to challenge an organism with an external stimulus might be met by an elaborate barrage of intertwined defence mechanisms as the victim tries to restore normality. Experiments are somewhat easier to define at the molecular level, but Paton is a self-termed 'good old-fashioned microscopist' who usually works with whole organisms; and unless one has elaborate equipment, it is difficult to be absolutely sure that there has not been cross-contamination between the plants one is testing and those that are kept to one side unmolested, to be used as controls, or that one is not being confused by a host of other complications.

For our part, we in the Venture Research Unit were highly sceptical at first, but the power and simplicity of Paton's argument, and his commitment over many years to his crusade, not to mention the opportunity we had of looking down his microscope and actually *seeing* whole plant cells packed with naked bacteria that convention insisted were not there, finally convinced us Paton was on to something very important. I knew that within normal plant cells there were organelles—the operating units of a cell—that look as if they have been descended from bacteria at some remote age in the past. The chloroplast, for example, is the organelle responsible for photosynthesis within the cell and another, the mitochondrion, controls energy generation, but both have their own DNA, which seems to be of a bacterial type. Perhaps in a long-past aeon, plants and bacteria had a closer relationship than they do now, and Paton had come across a remnant of that relationship. If he had, here, among many other things, was a new route for the introduction of novel genetic information into plants. But it would be *Nature's* way of experimenting with genes, and it might be more acceptable to the public than some of the trial-and-error approaches to genetic engineering that understandably have been the cause of much concern.

Paton, and his colleague from the Biochemistry Department, Anne Glover, were asking for funds that would allow them to put the work on a rigorous footing. Among other things, they needed to choose the species of bacteria and plants that would be the easiest to work with, and the protocols for the controlled inoculation of plants. Above all, however, bearing in mind the scepticism that abounded, they needed to devise a model system that would provide irrefutable proof that plants and bacteria in general, including their pathogens, could find ways of helping each other, and that we observers were not being deceived by some artefact.

I was not able, however, to convince BP to support a full-blooded attack, and we had to manage with modest funding for two years. The case I was able to make then was much stronger, and we got the green light we had all been waiting for. But the problems continued, and although progress was being made, it was tantalizingly short of providing proof in the form of a succinct experiment that would convince the sceptics. The evidence provided by rows of plants that were obviously doing well despite being pathologically challenged simply was not enough. For several years we shared the agonies of uncertainty and false starts, and some of my friends began to remark that it was not only the reputations of my Aberdeen colleagues that were on the line.

Major battles can rarely be won to order, and those we may have with Nature are no exception. Nowadays, however, research that does not regularly advance seldom survives, and scientists are therefore encouraged to concentrate on those more limited actions they believe they can win. The Aberdeen group were more fortunate than most, and our patience and their ingenuity were rewarded when they turned for help to a remarkable marine organism. Those who have been lucky enough to be at sea in a small boat on a calm, dark night will probably have noticed a soft phosphorescent glow coming from the vessel's rippling wake. This delightful display comes by courtesy of a company of marine organisms that include a bacterium, *Vibrio fischeri*, which as well as occasionally lightening our nautical darkness also indulges in a symbiotic relationship of its own. The deep-sea angler fish is equipped with a slender 'fishing-rod' that extends from its dorsal fin to dangle in front of its mouth. The tip of this eponymous organ offers a home to a population of *Vibrio*, and in return they provide a luminous lure to attract prey into the welcoming jaws of the angler. Since it is a messy eater, the microbes have a meal too.

The genes responsible for this sparkling performance, which not surprisingly have been given the name *lux*, code for the enzyme luciferase, which produces light as a by-product of its normal function—an oxidation reaction. The genes for luciferase can be isolated, and using genetic engineering techniques they can be introduced into the genetic machinery of any other bacterial cell. In its new surroundings, the luciferase enzyme emits light only if its host cell is active and can supply it with a ready source of energy. Thus, the stage was now set for a beautiful set of experiments. The group inserted the *lux* genes into the L-form of a bacterial pathogen, which in turn was associated with their test plants: French dwarf beans. A few days later, the plants were still growing, and using light-sensitive equipment in a darkroom, the researchers not only also saw the L-form bacteria glowing with contentment, but could watch the glow spreading as the plant grew. In a separate experiment, the wild, fully armoured pathogen, which had been equipped with *lux*, was used to infect plants that had earlier been treated with L-forms of the same pathogen, but without the *lux* genes. The group then watched the luminous glow gradually fade away and die as the pathogen's attack was frustrated, while the treated plants continued to thrive. Here at last was an experiment that, although it raised a flood of questions, should convince the sceptics that something very unusual was going on, and would also provide them with a sensitive indicator of a plant's health. Unfortunately, this breakthrough came shortly after BP's decision to close down the Venture Research Unit, as was foreshadowed in the Introductory chapter, and to which we will return briefly in Chapter 7. We have not yet been successful in attracting other sponsors for our crusade, but we will: good ideas will always prevail—eventually.

If Nature's complexity is infinite in all directions, there is little doubt that we understand only the tiniest fraction. Any walls we erect, therefore, around the bits we think we do understand may tend to hide from our eyes the vast extent of our ignorance, and also to obscure the fact that Nature's writ runs invariably through them all. As we have mentioned, the best scientists realize all this, but their views become diluted in the searches for consensus imposed by the funding agencies. Gulliver in his Travels met no more bewildering situations than the modern researcher has to deal with, and a latter-day Jonathan Swift would lack no amusement for his pen.

The notion of structure is one of Nature's themes, and the more we understand the better and more versatile our technology will be. It is easy, however, to see how even simple facts can become obscured by shifting our attention to the field of chemistry, which is the oldest experimental science, although for many centuries it went under the name of alchemy.

Those who have studied science, however cursorily, will have cut their scientific teeth by mixing two chemicals in a vessel—say a test-tube—to see what happens. After mixing, we are instructed to shake the test-tube to ensure that the reagents remain well mixed throughout the reaction. In industry, chemical engineers can exert their control only if stirring is vigorous enough to prevent dangerous local variations in concentration, a factor which often limits the vessel sizes and reaction rates that can safely be used.

Nature, on the other hand, is the most successful chemist we know, and Nature never, never does chemistry as we generally do it on Earth. (Martian chemistry is much more advanced than ours, but they have trouble with their physics. . . .) Neither man nor any other results of Nature's handiwork are well-stirred mixtures. Nature does chemistry at specific points in space as well as time, and in these respects Nature's chemistry is four-dimensional. Much of the chemistry we can control varies only with time; the spatial component generally has zero dimensions as engineers strive to ensure uniformity at every point in the reaction vessel.

Ilya Prigogene won the Nobel Prize in 1977 for his earlier work on the mathematical structure of systems far from equilibrium, which is another way of saying that they are rapidly changing. He discovered the rules by which so-called 'dissipative structures' are created in reacting systems. There are well-defined and stable regions in a reacting system whose composition is quite different from their surroundings, and are reminiscent of the many examples of self-organization we see in the natural world: in animals, plants, the oceans, the atmosphere, and indeed almost everywhere we look. However, dissipative structures had never been created by the hand of man, except transiently and sometimes by accident, simply because the reactions a chemist feeds with one hand are addictively stirred with the other, thereby obliterating any structures before they have a chance to grow.

This amazing story was first revealed to me by Harry Swinney during a visit to the physics department of the University of Texas at Austin. For some years, he had been studying the ways some types of chemical reactions apparently randomly change their behaviour over time: so-called temporal chaos. However, over long periods, an astute observer might notice that they are not random, but behave as if they are constantly being pulled towards a hidden object named a 'strange attractor' by David Ruelle and Floris Takens in 1971.

As might be expected, these patterns are invisible to ordinary eyes. Earlier, I hinted at the way Dudley Herschbach proposed to look at molecular behaviour using 'dimensional' spectacles. This unusual type of instrument is made by mathematicians who, of course, are exempt from the many constraints that must be suffered by opticians, and so if you wish they can supply a virtually infinite variety of aids to vision providing you are prepared to use your imagination as well as your eyes.

Just as spectacles transform our vision as we get older, the development of 'mathacles', as I would call them, has transformed our understanding in many areas of science. We might look at a pendulum, say, with 'mathacles' sensitive to velocity in one dimension and position in another, and we would immediately 'see' the pendulum bob moving round in an ellipse, rather than swinging from side to side as we normally do. These 'mathacles' would make it relatively easy to understand at a glance how friction and air resistance affect the motion. For other much more complex systems, the rate at which we will make progress will depend on selecting the right 'mathacles' for the job, or phase-space as it is more prosaically called in the theoretician's trade. The patterns shown in Fig. 2 have been produced using the phase-space composed of position in one dimension and the position of the same variable 53 seconds later in the other. These patterns were produced by Harry Swinney and his French collaborator Jean-Claude Roux in 1980, and indeed are the first recorded sightings of a strange attractor.

All this work had been done using the traditional 'continuously stirred tank reactor', or CSTRs as they are usually called, but now Harry Swinney wanted to use the understanding built up together with various collaborators to attempt to do chemistry in ways that might have Nature's approval. Chemistry is a tricky subject for physicists (and vice versa) and so he proposed to collaborate with a group of

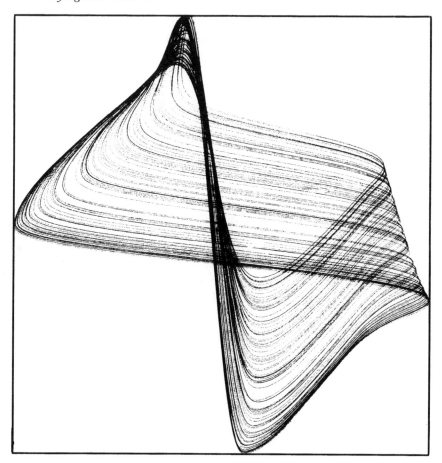

Fig. 2. The strange attractor, discovered by Roux and Swinney, constructed from measurements of the time-dependence of one of the species in a stirred chemical reactor using the 'mathacles' described in the text. The orbits that make up the strange attractor never repeat themselves. Roux called this beautiful shape the Texattractor, because they first saw it in Austin, Texas.

chemists: Jean-Claude Roux, Jacques Boissonade, and Patrick de Kepper from Bordeaux, and a theoretician, Werner Horsthemke, who at that time was also at Austin. This multinational group had worked together occasionally before, but never in such a concerted way.

When Nature does chemistry, she always arranges that precisely the right amount of reactants are delivered to the reaction sites at precisely the right time. If chemists are to surrender control of their reactions to Nature by throwing away their mixing machines, how are they to ensure that the right amounts are always available where and when they are required?

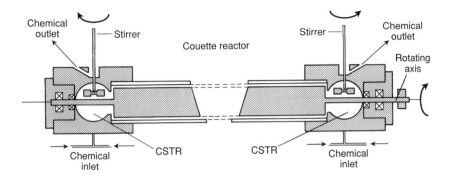

Fig. 3. A diagram of the Couette reactor. The continuously stirred tank reactors (CSTRs) are used to maintain constant chemical conditions at each end of the Couette reactor. They are fed through lines at the bottom, and the outflows are at the top near the shaft of the mechanical stirrers.

The Austin–Bordeaux solution was to let Nature do the stirring as well as the reacting. They proposed to use an apparatus invented by Maurice Couette in 1890, which scientists have found useful for studying the physical properties of fluids. The system consists of two concentric cylinders, with a thin annular space between them. If the inner cylinder is rotated, any liquid in the annulus is split into a large number of rings by the forces of hydrodynamics: the so-called Taylor vortices, in which mixing is usually complete. However, Harry Swinney now proposed for the first time to use the Couette system as a *reactor*, and thereby to add chemistry to the Couette repertoire. If conventional CSTRs are placed at each end of the Couette reactor as illustrated in Fig. 3, Nature then has all she needs to begin to do her chemistry, and Harry Swinney and his friends can watch her do it.

Unfortunately Nature would not be completely free because the Couette reactor gives only one degree of freedom, but the leap from zero dimensions had not been made before, other than transiently, and I could barely wait for them to start. The fact that half the group was in Texas while the other half was in Bordeaux seemed not to cause even the slightest difficulty. They planned to be in daily contact by phone and electronic mail ('E-mail' is the scientists' international communication system) and to visit each other regularly and exchange post-docs. Indeed, in my experience, when groups want to collaborate nothing will stop them: when they do not, nothing will make them. As a Hungarian scientist once remarked to me in the latter context: 'You can take a horse to water, but you can't make it swim on its back.'

The group had, of course, tried all the usual funding agencies, and although small amounts of money were available to support travel and good relations between the USA and France, they were proposing to cross too many divides to get the fairly substantial funds they needed for their campaign. Venture Research was delighted to help to open the door on the unexplored territory lying before them.

We have concentrated in this chapter on outlining some of the ways the existence of uncharted territories might be recognized. The three examples chosen have no special significance except they are from the natural sciences, and any three chosen from the current groups of Venture Researchers listed at the end of the book would have served our purposes. There is also no doubt that there are many more potential Venture Researchers we do not yet know about, and they all would have a fascinating story to tell.

It should also be realized that in emphasizing Venture Research, we have chosen a type of research that is a tiny portion of the tip of the scientific iceberg. Most scientists work within or close to one of the many mainstreams, and are engaged in either consolidating what we know, or extending the current boundaries of knowledge. These latter types of research are also essential to scientific progress, but they are intensely competitive, and predators abound. There are scientists commanding large groups who scan the literature for signs of new competitors reporting an interesting discovery. One boasted that in these circumstances a lone ranger entering his field would probably publish little more in it, for he would marshal his troops quickly enough to ensure that there was nothing left to do. If you are a scientist looking for a problem, therefore, you should search for the many types of wilderness that abound, because once you have found one there is a very good chance of being able to protect your pioneering position before it becomes a mainstream subject and the competition recognizes the importance of what you have done.

Nevertheless, there are many pitfalls for the unwary. Niccoló Machiavelli, almost 500 years ago, warned that those who would challenge the status quo have the most difficult time. Those who stand to gain by the change will not readily believe that you will succeed, and will offer only lukewarm support; conversely those who stand to lose their long-established privileged positions will oppose you with all the resources at their command. He was addressing his remarks to politicians, but our experience in Venture Research over the past decade or so has been that they apply equally to scientists now, and indeed they probably always did.

4

Playing with design

My own academic training was in physics, which according to convention is classed as a natural science, a term that unfortunately carries the pejorative implication that other types of science are somehow unnatural. Indeed, some engineers would prefer not to be dubbed scientists at all, for they say that it is a scientist's place to know, while theirs is to practise the art of doing. Even the word 'engineer', however, has its problems. It is drawn from the Latin word *ingenerare*, 'to create', but in the English language 'engineering' also acquired more sinister overtones in that it once meant to contrive, to manoeuvre, or to trick, and even today has not entirely lost these nuances. The relatively low status accorded to engineers, at least in England, perhaps has its origins in English customs and language, and in the academic disdain for practically oriented knowledge that persisted for so long. The universities of Cambridge and Oxford first established chairs in engineering in 1875 and 1903 respectively, considerably later than was generally the case in France and Germany, but it was not until well after the Second World War that the reputation for academic excellence that these ancient universities enjoy began to be associated with engineering too.

Yet the unknown person who invented and built the wheel was what today we would call an engineer, and it is astonishing that the formal study of the art that has so radically changed our lives has had such a chequered and brief history. I shall defer until Chapter 6 an outline of how some of these changes have been brought about: here we are concerned with research. It should also be made clear that there are those who do not agree with the simple distinction between science and engineering mentioned above, and I am firmly within that grouping. There is no question that engineers are doers, but how on earth can they *do* unless they also *know* a great deal about *what* they do? There seems no question, therefore, about whether engineers are scientists, except to ask what type of scientists they are.

Our response will be that in Herbert Simon's language they are, with others, scientists of the artificial, but although this somewhat pedantic point is nevertheless an important one, it is still not quite enough for our purposes in Venture Research. As has been outlined, we had decided that our quest should be to provide freedom to those few people who want to wander and wonder in the wilderness, to open up new fields, and generally to explore. We chose this policy as the best we could devise to restore diversity to the scientific enterprise, by which we mean *all* scientific enterprise, and to help precipitate those un-expected and inspirational discoveries that give zest to life, change the ways we think about some important facet of science, and eventually create new types of industry.

The snag is that sciences like engineering are about making things happen; not just anything but usually a highly specified and costed thing to be delivered by a definite date. Since research, and especially Venture Research, should include the possibility of a surprising out-come, there would seem to be a paradox here because engineering re-search is usually highly mission-oriented. An engineer commissioned to look at new designs of rocket engines for space travel would be expected to produce ideas that might lead to better rockets, and would expect little thanks if he or she were to come up with something com-pletely different, however inspired it might be. For engineering re-search, therefore, it would seem to be necessary to qualify the meaning of the word 'research' so that its target is always kept in mind; indeed, almost all engineering research seems to be qualified in this way. Choice of problem is usually the same as the choice of specific objective, and the research will consist of analysing, manipulating, and exploring the current understanding of materials and other resources, and the ways in which they may be used to achieve a particular end.

In some respects engineering is currently more of an art than a science in the sense that an artist can paint a picture that draws our attention and admiration, but will rarely be able to explain to the uninitiated how they too can wield brush and paint to weave magic on their own canvases. Such skills can apparently be acquired only after years of patient and prolonged practice, which would also seem to be the case in engineering.

Traditionally, the first major step at the beginning of a new engin-eering project is the drafting of a preliminary design. In most cases, the supervising engineer will draw on materials and resources generally available, on an extended experience of the ways they can be used, and the statutes and regulations governing their use. Occasionally, projects using state-of-the-art technology may stray into uncharted territory, whence the need may arise for a targeted research programme on the properties of materials, on control systems, on mathematics, on thermodynamics, on turbulence or drag, on catalysis or corrosion, on metabolic pathways, and indeed on virtually any subject on which disciplined enquiry is undertaken. The work may be carried out within an engineering company itself, or in the academic sector. The new work may lead to changes in the design and perhaps even more research before the supervising engineer is content that the information available is sufficient for the completion of a design.

Engineers are usually trained in all or most of these activities, and particularly in the evaluation of a design's fitness for its purpose. But there is one crucial area which at present receives little or no formal attention in an engineer's training, and that concerns the question of how an engineer should organize his or her thought-processes when deliberating on a design. Some might say that natural scientists receive no instruction in thought-processing either, but they are usually faced with a problem where there is little or no understanding of how Nature does something, and so their creative faculties are challenged in open-ended enquiry.

The creative leaps that are needed to reveal some new facet of Nature's behaviour can take the breath away, and we certainly understand very little of how we can train our minds to leap to order, so to speak. The design engineer is also required to be creative, but the circumstances are quite different, since there will usually be a time when the designer has enough information to hand on every material and resource involved to produce a viable design. That knowledge may not be perfect but the designer's task would in any case hardly be made easier if perfect information were available, since the most difficult problems are generated by the complex interrelationships between the materials and resources rather than by the isolated behaviour of any one of them, and by the additional complications that will be introduced when the design is used by people. The challenge to a designer's creativity is therefore to solve all these problems and to produce a design that in some important respect is better than the competition's.

How should this challenge be met? How should the engineer set about creating the procedures and protocols for bringing together all the reasonably well-described parts piled on an imaginary desk, and for combining them to form the final product to the customer's satisfaction? Typically, and ideally, the designer might consider all the contributing factors for minutes, days, or months, depending on the complexity of the problem, and at some significant moment pick up a pencil to sketch out a preliminary design. Thereafter, the broad outline of the project is probably determined, and the many other possible options might be closed consciously or sub-consciously, even though much later, under the highlights of hindsight, they might prove to have been important.

The engineer's formal training will have been of little help in all this, particularly if the proposed product would break new ground. In general, only those engineers who have extensive experience will be considered competent to create successful designs. Thus, after completing their formal academic training (even if as students they won all the glittering prizes), engineers will usually have to spend a further decade or two painfully learning how to practise the art of the possible before they will be deemed to be sufficiently qualified to take responsibility for major new projects or to act on their own initiative. In the natural sciences, however, it is not uncommon for young people to make outstanding contributions to their fields within a few years of graduation. The inexperience of youth alone, therefore, is not the problem.

These considerations, and many discussions with scientists and engineers from varying backgrounds, slowly gave rise to our strategy for engineering research, and indeed for any other type of research concerned with unravelling the complexity of human endeavours. If we could encourage or help stimulate an examination of the abstract processes that lead an engineer to propose a design, and particularly to eliminate sources of unnecessary complexity, we might be able to make that art more of a science that could be taught; and it might then perhaps be possible to reduce the long additional period of apprenticeship. Moreover, since these processes are abstract, they will most likely be applicable to all the many sub-disciplines of engineering— civil, mechanical, electrical, electronic, chemical, marine, etc.—and perhaps also to such disciplines as economics. If we could find those people who wanted to wander in this theoretical wilderness, who knows what the outcome might be? Thus, we had the prospect of engineering research programmes that would not necessarily have to be constrained by a specific and tangible objective, that might lead to new types of practical activity, and most of all, that might surprise us.

But who would be a *theoretical* engineer? As has already been stressed, we do not try to persuade scientists to work on problems set by others; rather we respond to those who are themselves thinking along exceptional lines and need freedom to bring their own ideas to fruition. Many of our industrial engineering colleagues, and particularly Oscar Roith, who was responsible for BP's engineering operations in the early 1980s, maintained their strong support for our strategy, but after ten years of an intensive series of visits and lectures to almost all the universities and most of what were called polytechnics in Britain, to thirty or so of the well-known North American colleges and universities, and to a few in mainland Europe too, we were not able to find an academic engineer, in the classical sense, trying radically to change the ways that people think about this important discipline, and therefore to break new ground.

The sciences of the artificial, however, cover a much wider ground than the classical engineering disciplines, and we were to receive a strong signal from a few people working in perhaps the most recently established fraternity of doers, namely that of computing science. Computing is as old as mathematics, and has long found practical application in commerce, in navigation, and in science and technology generally. For centuries, however, computing was laborious and error-prone, and the only assistance came from tomes of tables and logarithms, which while easing the burden somewhat were themselves liable to error as they too had been ground out by the same monotonous and manual means that the poor wretch looking for help was obliged to use.

The first tentative attempts to automate computation were made in the seventeenth century, notably by Blaise Pascal, and later by Gottfried Wilhelm Leibnitz, who built a more ambitious machine that, for example, could calculate square roots. None of these machines, however, was reliable, and they hardly threatened the ancient abacus, which must qualify as one of mankind's most successful inventions. (During a recent visit to Tokyo, I was suprised to find this simple device still in daily use in some of Tokyo's shops, apparently in preference to the high-powered low-cost electronic wizardry that is now widely available.)

This situation began to change during the 1820s when Charles Babbage, the Cambridge mathematician, began to invest his considerable personal fortune and the rest of his life in trying to produce an 'analytical engine', and to reform nineteenth-century British attitudes to science and technology. Unfortunately, he failed in both quests, but the designs and prototypes he produced for his highly sophisticated machine proved long after his death to be seminal, and made major contributions to the development of computing during the 1930s and 40s.

Punched cards had been used by Joseph-Marie Jacquard to control (or, in the modern language to program) his weaving loom of 1805, and this idea had also been borrowed by Babbage. But Herman Hollerith, the US statistician, used it to remarkable effect in 1887 when he developed a card-reading machine that was able to process the 1890 US census results in three years, a remarkable feat at that time. This machine was later developed for commercial use, and eventually it laid the foundations for the emergence of one of the world's largest corporations: IBM.

War has often been a watershed in the emergence of technology, and this was certainly the case for computing. One of the most urgent preoccupations in time of war, hot or cold, is the protection of so-called intelligence during these prolonged periods of madness when secret information has to be exchanged. The usual way of achieving this end is to transform the characters that make up the signal in a carefully specified way known as a key. The most secure way is to arrange that each key is chosen from a set of keys of sufficient complexity to baffle the would-be code-breaker, and that each key is used only once. Thus, for example, sender and recipient might be issued with identical so-called 'one-time' pads, each containing many pages of randomly generated numbers. A typical signal might be preceded by a number, say 1234, which would tell the recipient to go to page 12, column 3, line 4 in the corresponding pad. Messages are often coded as groups of four numbers, so the key would be added to each group before transmission by radio, say, to be subtracted later by the receiver. Page 12 of the one-time pad in this example would then be destroyed. This technique, however, requires prodigious feats of organization and distribution, and is not always practicable.

A possible alternative is to use books of keys. They are much less secure because they might be captured, and although one-time pads may also fall into the wrong hands, they should contain only keys that have *not* yet been used, whereas the longer that books of codes have been in service, the more likely it is that its keys will be used more than once. This is a risk because typical messages contain a number of common phrases used repeatedly. Messages might all begin with say . . . to the Commanding Officer . . . , refer to . . . Monday 05.30 . . . weather unsettled . . . etc., and contain shorthand versions of common expressions or sayings, the best known of which is perhaps 'snafu', the acronym for—situation normal, all fouled up—to indicate the usual utter confusion of a war zone, and an observant cryptanalyst might recognize the recurring patterns.

The German forces decided to use an electromechanical machine called 'Enigma' to encrypt their messages. Enigma had been developed during the 1920s as a commercial machine, and looked something like an old-style cash register. In fact it was a highly sophisticated combination of spinning rollers and electrical switches, and the much more complicated war-time machines could produce between some 10^{21} to 10^{23} different codes according to the type of machine being used. (Each of the armed forces had its own version.) Small wonder therefore, that the Germans thought their system was absolutely secure, and the likelihood that sooner or later an Enigma machine would fall into Allied hands apparently gave them little cause for concern. They firmly believed that the only possible way to decode an Enigma message was to have the correct type of machine *together with* a knowledge of the precise initial settings used to generate that specific code.

The British, however, had two extraordinary strokes of luck. The first came from the Polish resistance movement and a small group of Polish cryptanalysts, who before the war started had not only been able to deduce Enigma's design, but also to produce a copy of the Enigma machine being used at that time. This *tour de force* was invaluable in that it gave the British the structure of the early, relatively simple German codes, but with the introduction of the more complex wartime machines there was still the little matter of discovering which of the 10^{21} or more settings had been used for a particular message. The second stroke of luck came in the shape of a young Cambridge mathematician called Alan Turing.

Turing might be described in today's language as odd. His biographer Andrew Hodges describes him variously as awkward and untidy, shy and unworldly, excitable and enthusiastic, and a sparkling scourge of bureaucratic authority. He was also in the language of the 1940s, '. . . a mathematical Flash Gordon and a logical superboy'. This inspirational individual graduated in mathematics at Cambridge, and took a Ph.D. at Princeton. On his return to England in 1938, he joined the Government Code and Cypher School at the age of 26. The School was known as GC and CS according to the fashion for acronyms, but also as the Golf Club and Chess Society to its members. It was then located in London, but at the outbreak of The Second World War it moved to Bletchley Park in the Buckinghamshire countryside, which Hodges describes as being at 'the geometric centre of intellectual England' because of its roughly central location between Oxford and Cambridge. The School's analysts, largely linguists and classicists, had struggled with Enigma without much success for some years before Turing arrived, and they were not optimistic, but in view of the extreme importance of the task the British Government had decided to broaden the base of the School's expertise to include mathematicians in particular. Turing was one of the new recruits, and one of the School's first mathematicians.

Mathematics is unique among the sciences. We have already quoted Roger Bacon's famous aphorism, 'Mathematics is the key and the door to the sciences', but sweeping and all-embracing as this statement seems to be, it nevertheless fails to encapsulate the essential quality that sets mathematics apart. After all, physicists or chemists among others can and frequently do make similar claims. Mathematics, however, is the only science that can be wholly intellectual. Scientists can wander in the mathematical wildernesses and make major discoveries without having to make recourse to a higher authority such as Nature or any worldly institution as other scientists must, save that which is found in men's minds. The mathematical universe, so to speak, is limited only by the imagination, and extends from the infinitesimal to the infinite over a landscape that seems to show structure of boundless variety no matter which way round one's mathematical telescope is used or how powerful it may be.

There must have been little doubt in Turing's mind therefore that mathematics was the key to Enigma, for what was 10^{21} but a simple number? One of Turing's interests was numbers. To those not versed in the science of arithmetic, it might not seem very interesting. One can write down a number, a page of numbers, or a book of them, a set of tasks that would seem to be limited only by one's patience, paper, and pen. But numbers exist in distinct domains, each of which has its structure and hierarchy. The natural numbers, 0, 1, 2, 3, etc., are perfectly precise. Some have factors, as does the number 10; some do not, $10 + 1$, say, that is they are prime. But if $10 + 1$ is prime, is $10^{10} + 1$, or $10^{100} + 1$, etc., also prime? Since the prime numbers have a special significance even in today's codes, commercial and military, how should the questions be answered? Rational numbers can be expressed as ratios of the integers, $1/2$, $3/4$, $5/6$, etc., but some numbers, $\sqrt{2}$ or π for example are irrational, and cannot be expressed as a ratio. Both $\sqrt{2}$ and π, however, can be expressed as a series of ratios, but each series is infinitely long, and although these numbers can be calculated to arbitrary precision, they cannot be known exactly. Can all irrational numbers be expressed as a series? Which numbers can be calculated? Turing had devoted his postgraduate work at Cambridge to questions such as these, and so was well prepared for the formidable challenges that lay ahead.

During the war, the German authorities changed the rotor settings for Enigma each day, and the School's task therefore, was to construct a machine that could deduce the daily key from the coded messages. Their first machine or single-purpose computer as we might call it today, consisted of high-speed relays and rotors of the type that were once used in telephone switching systems. These 'Bombes' as they were called, because of the insistent ticking noise they made, were formidable machines, whirling and clicking away night and day as they sought to identify the key, that is the one combination for each day's signals that would lead to a logically consistent result, and a set of comprehensible messages. They had to be programmed, of course. Luckily for the Allies, the first six months of the war saw little action—this was the period of the 'phoney war', as it was called—and so the School had some respite. Turing drew inspiration from his colleagues, who worked closely as a team, but he is credited for what turned out to be a crucial discovery without which their task would probably have been impossible.

Computer programs are based on algorithms: sets of procedures or protocols that a program will use. Turing's crucial step in designing the algorithm for the Bombes was based on a subtle and complex symmetry argument. Something of its flavour can be drawn from the idea that if the correct relationship between any two letters in a coded and decoded message had been deduced, all the other relationships for any one letter, that is 25 of them for a 26 letter alphabet, would be incorrect. His much more profound observation reduced the possibilities they would have to examine by a factor of 26, and just brought their task into the realms of the possible and practicable.

The Bletchley Park team must have been one of the largest collections of brilliant people ever brought together to tackle an intellectual problem. They revelled in reasoning and logic as they struggled to reveal contradictions and inconsistencies that might help to reduce the formidable numbers of possible code combinations that the Bombes would have to examine as they clicked around the clock. The German Air Force codes seemed to be the easiest to crack, and were indeed broken during the spring of 1940. Even then, it might take from two to twelve hours to reveal a day's key (and sometimes much longer) and to release the consequent cascade of that day's messages. It was ironic that the German Air Force were tuning up their Enigma arrangements at the same time, and the first decoded messages turned out to be nursery rhymes used as practice.

Decoding messages was one thing; interpreting them was another. As Sir Harry Hinsley and others report in their voluminous account of 'British Intelligence in the Second World War', GC and CS decoded on 11 November 1940 a message of 9 November apparently describing a proposed major bombing raid by the German Air Force. The code name for the attack was 'Moonlight Sonata' and there was a mysterious reference to 'Korn' which might have been a code name for the target, but there was nothing to indicate that it was. No precise date for the attack was given, but a period of full moon began on 14 November. Eventually, information from a prisoner of war shot down on 9 November, and other Enigma intercepts, pointed to Coventry, but at this relatively early stage of the shooting war, there was some confusion, to say the least. The countermeasures taken to offset what turned out to be a colossal raid on 14 November by some 500 German aircraft were largely ineffective, but among many other things, the subsequent flood of precise and priceless information from Bletchley (an average of 3000 decrypts a day) was to help forge unprecedented degrees of cooperation and delegation among the Allied Forces, and eventually to direct connection with operational forces when circumstances demanded it.

The breaking of the German naval codes in early summer of 1941 was most crucial, and was an operation for which Turing had overall responsibility. U-boat supply ships were located and systematically destroyed, and as far as possible convoys were routed to avoid known U-boat locations. Bletchley Park's contribution to the Battle of the Atlantic was perhaps their single most important contribution. Later, in the early autumn of 1941, the German army codes were also broken, and there was hardly a theatre of war in the west that was not overseen by the intellectual warriors of Bletchley. During a visit to Bletchley that summer, Winston Churchill referred to them as 'the geese who laid the golden eggs and never cackled!' Hodges went on to remark 'Turing was the prize goose!'

During 1942, routine intercepts had revealed a small amount of traffic entirely different from Enigma. Instead of being transmitted in Morse Code, the signals were teleprinted, that is the characters were represented by holes printed in paper tape. Turing also played a valuable part in breaking this code, usually referred to as 'Fish', which was eventually found to be a channel of communication between the most senior levels of the German High Command. A new type of machine was needed, of course, because the purpose-built Bombes could only tackle the Enigma codes. Thus, the Colossus was conceived, a computer that used some 1500 thermionic valves. Brian Randell, writing in *A history of computing in the twentieth century*, relates that Colossus seems to have been the first programmable electronic computer. It is almost certain that so many thermionic valves, which are notoriously unreliable, had not been put into one machine before, and the machine's success was no small tribute to the people who designed it, and to those responsible for its continuing operation around the clock, no doubt always under the gun, and plagued by the constant anxiety that people's lives were hanging in the balance. However, they must have worked in an atmosphere that was stimulating to the point of euphoria: they knew the importance of their work in the most general terms, and the impact it was having. As the war progressed, they were also helped considerably by the highest priority that was given to keeping the School supplied with equipment, no doubt to the bewilderment and envy of those who were not in the know. Randell relates the story of an engineer telephoning the Ministry of Supply for another couple of thousand valves and being asked, 'What the bloody hell are you doing with these things? Shooting them at the Jerries?'

The work on Enigma was code-name 'Ultra', and was perhaps *the* most closely guarded secret of the war and for almost thirty years after. Everyone knows of the Battles of Britain, of Alamein, of D-Day, and others, but at Bletchley Park a battle of supreme importance was fought which apart from those directly involved was known only to a tiny and very, very exclusive few. Never in the history of human conflict had such a war been fought. A wholly intellectual war, a war of wits, reason, and logic, an entirely defensive war in which no physical casualties were suffered. The Second World War was prolonged and bloody, and millions of men and women gave their lives to bring about an outcome generally regarded as just. But even with the enormous advantage of Ultra, it was a close-run thing. If the Enigma codes had not been broken it is highly likely that the balance would have shifted, and among many other possibilities nuclear weapons might have also been used in the European theatre. One might say that in this respect at least, reason prevailed.

Group Captain Frederick Winterbotham was one of the promoters of Ultra, and was responsible for the coordination of the Ultra output among the Allied forces. After Pearl Harbour, and the entry of the United States into the war in December 1941, Ultra information was shared with the Americans, who of course had their own considerable cryptanalytic capability. However, at the end of the war, Winterbotham reports in his book *The Ultra Secret* that the Supreme Commander of the Allied forces, General Dwight D. Eisenhower, who later became President of the USA, expressed his view that 'Ultra was decisive'.

The German High Command knew that their secrets were being regularly penetrated, but it seems that their faith in Enigma machines was absolute, and they attributed their problems to human fallibility and espionage. I am reminded of Rudyard Kipling's lines written long before the war (he died in 1936) entitled 'The Secret of the Machines':

> But remember, please, the Law by which we live,
> We are not built to comprehend a lie,
> We can neither love nor pity nor forgive,
> If you make a slip in handling us you die!

This phenomenal success story was known only to a few for decades, and it is perhaps a consequence of the profound secrecy with which Ultra was shrouded that computing in Britain somewhat lacked for stimulation once hostilities had ceased. Its focus, however, was on the problem of how the successors to Colossus should be built. Turing worked on ACE, the Automatic Computing Engine, thus continuing the Babbage tradition, while across the Atlantic another towering intellect, John von Neumann, was working on the ENIAC, the Electronic Numerical Integrator And Calculator, a machine that had been designed by J. Presper Ekert and John W. Mauchly in the engineering department of the University of Pennsylvania. Von Neumann had a wide range of other interests too, and during the war had worked on the Atomic Bomb, the Manhattan project.

One of the justifications offered to support the building of the proposed ACE was that it could be used for ballistic calculations, but Turing's vision was for a *Universal* Computing Engine. Turing neither liked the slow pace imposed by post-war bureaucracy nor the steady dissolution of his dreams as the ACE design was watered down. A pilot ACE was built in 1951 at the National Physical Laboratory near London, where Turing had moved after the war, but by 1948 he had moved on to the University of Manchester, where they were also building a machine. The ENIAC was already working in 1945, but was too late to help the war effort. It was designed without any knowledge of the Colossus, of course, and understandably, therefore ENIAC was widely held to be the first electronic computer even though the highly secret Colossus predated it. ENIAC was certainly the first in a long line of progressively more ambitious and successful machines, and Turing might well have regretted his decision not to accept von Neumann's offer of an invitation to work with him while he was at Princeton.

Tragically, Turing died in 1954. He committed suicide two years after being found guilty of a homosexual act, which was a criminal offence in England at that time.

Turing and von Neumann, however, are generally accepted as the two pioneering fathers of what is now called computing science, a term that came to be accepted only during the 1950s and 60s. Until that time the emphasis had been on hardware, that is the equipment, the thermionic valves, capacitors, and resistors, etc., and the panache and flair of the engineers who kept them operational. The discipline of programming, however, that is the consideration of *how* a machine might be used, rather than *what* it might be used for, tended to be secondary, a situation that must have perplexed Turing since it was precisely his attention to these intellectual considerations that had resulted in Ultra being a secret worth keeping.

Shortly, however, a discovery was to be made at the Bell Laboratories that would transform the capabilities of computers perhaps beyond even what Turing or von Neumann could imagine. In 1947, John Bardeen, Walter Brattain, and William Shockley discovered that some types of electrically insulating materials like pure silicon or germanium are given remarkable electronic properties when minute (parts per million or less) quantities of impurities that can donate or accept electrons, arsenic or aluminium say, are combined with them. Thus transistors were born, which shortly would lead to developments that would surprise even Bardeen and his collaborators.

The early transistors were used in some respects like miniature thermionic valves, but they were a tenth or less than the size of the brightly glowing glass tubes they replaced. Transistors are made by allowing trace quantities of the chosen impurity to diffuse into the parent material. It was soon found that this process could be controlled with such precision that a complete electronic circuit could be 'written' on to the parent material using techniques originally borrowed from the printing industry but which were subsequently so refined and extended that a full page of a newspaper could be 'printed' within the space enclosed by one of that page's full stops. Newsprint is passive, of course, but this new type of lithography can produce arrays of microscopic semiconductors—integrated circuits—of phenomenal power. Nowadays there is scarcely a facet of life in an advanced country that has not been either touched or transformed by the application of these techniques and by computers in particular.

Passive though the printed word might be, the pen has long been regarded as mightier than the sword, but only to the extent that it has been pushed by a person who profoundly understands the issues at stake. For the computer programmer, the satisfaction of this condition is a routine and rigorous requirement. Typically, if a programmer wants a computer to do something, he or she must assemble all the information the computer will require, neither more nor less, and divide it into portions, each of which the computer will accept, like a coin in a slot, to initiate one of its functions. Each portion must be in perfect harmony with all the others, and together the sum of the functions so commanded must result in the performance of the required task. Today's computers operate at speeds ranging from millions to thousands of millions of calculations per second. Many calculations might be required to carry out a specific function, but the essence of the programmer's problem stems from the fact that the quality of the program might be put to the test every time a calculation is performed. Those who would program must therefore be prepared to allow their thoughts to be subject to a relentless torrent of evaluation, and must know that if their work contains the subtlest flaw or inconsistency, commonly called a 'bug', failure will result.

During the 1940s and 50s, when computers were relatively small, the overwhelming emphasis had been on the hardware, as has been mentioned. If a program failed for some reason, the bug could usually be identified and little damage would be done. Progressively, however, the balance shifted. The cost of producing hardware fell and its reliability increased to the current levels of near-infallibility. The use of computers became increasingly ambitious and safety-critical, and the emphasis shifted to the production of efficient and reliable computer programs and routines, that is to the software.

In 1980 this seemed to be a fertile field to search for Venture Researchers, and following a meeting with Professor Christopher Longuett-Higgins at Sussex University, I went in November of that year to see Professor Edsger W. Dijkstra at the Eindhoven University of Technology in the Netherlands. He is a professor of mathematics, but as we settled down in his home at Nuenen, not far from the University, to drink coffee with his wife Ria, it became clear that the mathematics he professed might just as accurately be called a discipline of reasoning. My own training as a physicist had necessarily entailed the acquisition of a range of mathematical skills, but the mathematics Dijkstra patiently revealed was starkly different from anything I had thought about before. He was not concerned with mathematical results—theorems, equations, techniques, etc.—with which most mathematicians and students of mathematics are preoccupied, but with *how* to do mathematics, and particularly with how to organize one's intellectual efforts so that complexity can be recognized, tamed and brought under control, and man-made complexity can be avoided or eliminated.

Here was wilderness indeed! Here was a crusader who not only wanted to change the ways we think, but wished to do so in ways that might expose concerns of which we were yet unaware or had not recognized as having importance. The number zero might serve as a simple example. How many of us consider zero to be a number, that is a fully paid-up number like one, two, or three? The number one is usually defined as the lowest of the natural numbers; in the west, at least, our education and culture result in the deeply ingrained belief that zero is a special case, a null event, or an absence of number. But in mathematics, and in computing science, failure to recognize the numerical status of zero results in the generation of unnecessary complexity. I remember having difficulty at primary school with the following simple problem: Four nails are driven into a piece of wood in a straight line. Each nail is placed one foot from its neighbour. What is the distance between the outermost nails? The answer, of course, is not four feet. In problems of this type we were told we had to subtract one. But if the first nail is labelled 'zero', the last will be labelled 'three' and there is no need to remember any other information, or to draw a diagram; in fact, had we been taught to count from zero at the outset, the 'problem' would become so trivial that teachers who like to set such traps would have to think of something else.

Complexity has many origins, of course. Once upon a time, long ago, the board of a railway company in its determination to cut costs decided that in future only half of its passenger carriages would be fitted with a toilet. Unfortunately, this decision was not conveyed to the yard where all trains were assembled. They continued to handle carriages in the same way as before, and some trains left the yard with hardly any toilets at all. In the face of complaints, this lavatorial omission was rectified by labelling carriages according to whether they had or did not have a toilet, and instructing the yard to ensure that the numbers of carriages of both types were roughly equal. It was a complication, but the yard's shunters were soon complimenting themselves that it was one they could manage.

But the complaints continued, and it was soon discovered that although half the carriages of each train were blessed with a toilet, they were sometimes bunched together at either the front or the rear. The yard was therefore issued with new instructions that carriages with and without a toilet should alternate. The yard were not happy, but they complied.

The passengers' complaints were hardly stemmed, however, because carriages with toilets had them only at one end, and a tired traveller turning the wrong way might sometimes have to struggle through three carriages before finding relief. The answer, of course, was to instruct the yard to place all toileted carriages the same way round, but no sooner had the long-suffering shunters recovered from this latest signal than new orders had to be issued because inconvenienced passengers did not know which way to turn in their quest for comfort. It was therefore ruled that henceforth carriages had to carry arrows pointing clearly in the direction of the nearest toilet, which meant that every carriage was now to be turned into a directed object. Revolt ensued. The trains could not be assembled within the allotted time, the shunters said, and the board were worried that the increased costs would wash away the original savings.

At that moment, a passing philosopher pointed out that the board's problems could be solved by starting afresh with new orders for the yard. Henceforth, each toileted carriage should be permanently joined at its toileted end by a non-toileted carriage, and the yard could shunt these double units any way they liked. Every train would now have an even number of carriages, but that was acceptable to the board; so everyone, as Dijkstra concludes his story, lived happily ever after, and flushed by success, the world's first competent programmer had taken a bow.

The Dijkstra philosophy is drawn from a full recognition of mathematics as a wholly intellectual discipline. Thus, the primary activity of a mathematician should be to *think* about a problem, to reason about which of its facets—he uses the term 'concerns'—are relevant and important to a solution, and which might be ignored or subsumed into other concerns. The overwhelming urge that most of us feel to write things down should be deferred as long as possible, because writing gives thought tangible existence, and a premature commitment closes down options subconsciously that may be valuable later. When the structure of a solution is clear in our minds, each step of the argument should finally be set down on paper in our best handwriting as a fitting tribute to the elegant economy of our reasoning.

The relevance of all this to computing science is rooted in the recognition of the relationships between programs and mathematical proofs. The proof ought to be a succinct summary of the justification of a proposition, whereas a program is a series of statements that should cause a computer to perform whatever task a programmer proposes.

The growth of the discipline of programming as the poor relation, so to speak, of the glamorous sciences underpinning the phenomenal growth of computer power has meant that a programmer generally takes an experimental approach to the task of programming, and produces as soon as possible what seems to be the best that can be done, *with the expectation that the first attempt will probably fail.* Thereafter the computer's diagnostic capabilities can probably be used to help reveal the bugs. After a few iterations, the program will probably run well enough: most if not all the bugs will have been revealed, and a point of diminishing returns will have been reached in the search for any that might remain. The program might then be accepted, even though there could be no indication of whether a prospective failure would be of minor or major importance.

However, as Dijkstra would say, this form of program-checking can reveal only the presence of bugs and not their absence. On the other hand, if the programs are written as if they were mathematical proofs, their correctness is guaranteed, and there will be no bugs. Thus, the mathematician–programmer's task is to develop the guidelines that lead to the recognition of effective arguments, and to avoid unnecessary complexity. Programs prepared in this way descend a logical staircase of even pitch, each statement inevitably following the preceding one, without having to resort to clever tricks or pulling rabbits out of a hat.

At present, this is easier said than done, and one purpose of Dijkstra's research is to uncover the types of problem that are amenable to rigorous treatment, and progressively to expand their compass. Nowadays, computers are being used increasingly in safety-critical situations—air traffic control, railway signalling, chemical plants, etc.—but there has been a need for precise software for decades. In October 1969, during a pool-side interlude at a conference in Rome, Dijkstra asked Joel D. Aron, who had been the IBM scientist responsible for producing the software used to control the lunar landing from the *Apollo 11* flight of the preceding July, how he had ensured that the software was correct. The response was that they had not been able to do so. Indeed, Aron went on, they discovered quite by accident only five days before the *Apollo* lift-off that the Moon's gravitational force had been given the wrong sign; that is the Lunar Module's software had been written to control a landing on a planet with a *repulsive* rather than an attractive gravitational field. 'They were lucky', Dijkstra wryly remarked.

Every computing scientist has a fund of horror stories relating to the ways in which the most trivial errors, such as the omission or unnecessary inclusion of a comma, can lead to the most horrendous consequences. It is not so much that Dijkstra was saying that if only people were clever enough they would not make mistakes, but that if, like golfers or cricketers who develop bad habits like lifting their heads before playing the ball, programmers use unsatisfactory notations and structures accidents will inevitably happen; the only uncertainties will be when, and how severe they will be. The widespread use of the word 'bug', for example, instead of the more accurate 'error', seems to imply that most programmers believe that the probity of their programs is somehow beyond their control, since even the most health-conscious people can after all become infected.

For his part, Dijkstra practises what he humorously and patiently preaches. The programs or algorithms he derives are either numbered and circulated by mail (the parable about shunting is labelled EWD 594) or published in the literature, but they are never tested beforehand by running them on a computer. Parables apart, he has no need to, because they have been written in a format that virtually ensures their correctness. It does not mean, necessarily, that they are perfect. Subsequent discovery of, say, a new symmetry argument may improve the efficiency of a mathematical argument, but without affecting its veracity; the equation : $1 + 1 - 1 + 1 = 2$, for example, is just as valid and correct as $1 + 1 = 2$.

We did not take long to make up our minds, and in January 1981 Dijkstra became our first Venture Researcher. He hardly needed freedom, however, to bring his ideas to fruition, for he had been doing precisely what he wanted to do for many years before we met, since his research requirements largely consisted of time, pencil, and paper, and perhaps a photocopier so that he could share his work with his colleagues. But Venture Research ideals seemed to appeal to him, particularly those that might increase industry's interest in what he was doing and trying to do. In any case, a respected colleague, Netty van Gasteren, had run out of support, and needed a salary and space for a desk. The Eindhoven University of Technology agreed, and we were very happy to be associated with their crusade, which continued when Dijkstra moved to the University of Texas at Austin in 1984. We still are.

Other computational pioneers soon joined the Venture Research fold, notably Professor Rod Burstall, Dr Gordon Plotkin, and Dr Robin Miller (both now professors) at the University of Edinburgh, and Professor Robert Boyer and Professor J Moore at the University of Texas at Austin. Our patron BP gained great benefit from a series of intensive interactions with these inspirational individuals during the 1980s that gave new strength and structure to the company's considerable interests in computers and computing.

The noticeable lack of mainstream engineering interest in Venture Research was, however, a source of concern. That has not changed over the years. It is perhaps hardly surprising, since, although the intellectual content of any engineering enterprise is considerable, the overwhelming emphasis in the profession is on the objectives to be achieved rather than on *how* they might generally be brought about. Theoretical engineers might therefore have been expected to be thin on the ground. Although we do not question that engineers must be doers, in Venture Research we are looking for those whose natural inclination is to subordinate their 'doing' capabilities to thinking about new ways in which things in general might get done.

It was perhaps remarkable that our first (and only) theoretical engineer from the mainstream was to emerge from the industrial sector. Graham Parkhouse advised BP between 1979 and 1981 on the use of composite materials made in a novel latticed form. At that time, he was a senior design engineer with the W.S. Atkins Group, a large British firm of engineering consultants for which he had worked for some twenty years, and so one might say that his engineering apprenticeship had been well and truly served. During his work for BP he knew that there was no way of predicting the behaviour of complex material composites: the only way forward was to draw on his company's extensive experience of structural analysis and materials in general. He also realized, however, that this was a symptom of a more pervasive problem in that there was no way of quantifying structural performance.

All structures are, of course, made of something, and the performance of the classical materials such as iron, steel, concrete, and wood, their stress–strain relationships, density, cost, etc., have been extensively documented. Prosaically speaking, people too are structures, but there are few who care about what say Marilyn Monroe or Arnold Schwarzenegger might be made of. For engineered structures, it is indeed *shape* that is designed: the geometry or topology or configuration that describes how the material is to be used. Although shape and material are entirely different entities, when it comes to engineered structures they seem to be inseparably related in designers' minds.

A structural designer will typically ensure that the stresses imposed on a material are always less than the permissible stresses laid down in the standards and codes of practice for the use of that material. But structural failure is not synonymous with material failure. Structures can fail by buckling without any material failing; conversely, materials can fail, or corrode, or be damaged, but because the design allows for a large margin of error, the structure can still be fit for its purpose. Permissible stress criteria for the use of materials are generally derived from extensive experience, and can depend on the type of structure in which they will be used. The barriers to the introduction of new materials or structural forms can therefore be considerable, simply because the materials in question are untried. Structural design proceeds by a series of iterative steps in which an outline shape is sketched, followed by a rigorous analysis of the performance of each material component used to give that shape, which in turn may suggest changes to the original design. What Parkhouse wanted to do was, in Dijkstra's language, to separate consideration of the 'concerns' of shape on the one hand, from those of material on the other, and help to put more science into the design process.

Trying to think about shape in an entirely abstract way brought back memories of the childhood game of trying to think about a horse without also thinking of its tail. Changing the ways in which we think is always very difficult, but Parkhouse had an ingenious approach. Say that you have been asked to design a bridge to cross a well-surveyed river. The next step would normally be to sketch an outline design of a specific type of bridge—suspension, cantilever, etc. This will almost certainly mean that consciously or otherwise you will have chosen the materials too. What Parkhouse was suggesting was that we should take a hierarchical approach to decision-making; if a small group of people want to play a game, the hierarchical way to decide precisely which game they should play might be to ask: indoors or outdoors? Say every-one agrees indoors; which game? Say cards; which card game? Say poker; which poker variation? Say stud. If, on the other hand, stud poker had been suggested at the outset someone might have said they preferred golf, and no progress would have been made. For structures, Parkhouse was suggesting that we should start with the largest and simplest overall shape and progressively descend through a few hier-archical levels until we are considering nuts and bolts, or nails, etc.

For our bridge, the simplest overall shape might be a lengthy tube or rectangular box filled with an unspecified mixture of materials and space that might be thought of as a virtual material of low density. The design process would then be to suggest an overall shape for the bridge, as might normally be done. Shape is hierarchical. Thus, bars can be

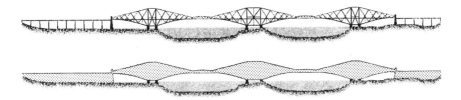

Fig. 4. The Forth Railway Bridge, built in the 1880s, is a classic example of a hier-archical structure. The members making up the main lattice are themselves latticed, but on too small a scale to be seen in this illustration. The blurring of the discrete lattice members into continuous grey material infill (shown in the lower drawing) illustrates the way that Parkhouse proposes we should consider structural form during design. The material filling out the envelope can be any number of configurations of materials dispersed in different forms. The performance of each configuration is quantifiable, so that the choice of infilling can be an informed one.

latticed to make a jib for a crane for example, and in turn jib-like structures can be latticed to form a structure like the Forth Railway Bridge, as can be seen from Fig. 4. At each hierarchical stage, the material—steel in the case of the Forth Bridge—is distributed more sparsely, and the average material density drops at each successive stage. Parkhouse was proposing that designers could draw on a library of the static and dynamic material-transforming properties of the many forms and then select the combinations of form and real materials that best met the specification and could be combined to fill out the overall shape.

It would be expected that this new approach would be unlikely to lead to new ways of using traditional materials like steel and concrete, since all the pitfalls have long since been well and truly explored, and designers can usually turn the powerful beams of hindsight to illumin-ate almost every conceivable problem. But Parkhouse's methods might reveal new reasoning on old structures. At the very least this might be a stimulating teaching aid. What was more exciting was the prospect that these methods might help with the use of composite and other high-performance materials such as carbon fibres, where gaping chasms of uncertainty await to extract their expensive exploration fees.

The response of my colleagues at BP was interesting. At the middle-ranking levels the engineers tended to be dismissive and critical, whereas their scientific peers tended to be excited and enthusiastic, as also were the most senior engineers and scientists. It is usually the case that the engineers with direct and immediate responsibility for projects tend to be the less senior, and if things go wrong the sky usually falls first on them. It is hardly surprising therefore that those in the firing line might be suspicious of someone pointing to brave new worlds just as they might at last be coming to grips with the old. It would be their careers that would suffer if the new ways led to disaster, whereas if they were successful the biggest prizes would tend to go to those who gave the directions. The natural scientists would tend to applaud work that might increase an engineer's predictive capability, whereas the senior engineers, having been through the mill themselves, could see the need for new intellectual tools to meet the challenge of new materials.

For my part, I had not previously met a practising engineer of such enthusiasm, insight, and vision. Nevertheless, it was more than six months before the significance of his thinking finally sank in, during which time Parkhouse's ideas changed too, as they matured and blossomed in the critical but encouraging environment that is part of Venture Research. In his case there was another problem in that he could hardly enter into a free-wheeling programme of research within an industrial environment. He would have to resign his post at Atkins. We undertook to help find a university not too far from his home that would give him the space he needed. It was not easy. The University in question (Surrey) needed to be convinced that someone with more than twenty years successful experience in one of the country's biggest design consultancies was a suitable person to occupy an academic post, even though he would not be required to teach; and in BP there were those who were worried that we were turning Parkhouse into a gypsy who might have to wander for the rest of his career. We prevailed, however. Parkhouse was fully aware of the risks he was taking and we were glad to have launched him on his exciting new career.

5

Building an academic career

The previous three chapters have outlined some of the factors that might bear on a newly qualified scientist's choice of a research topic. These problems are not confined to the young, who after all have little experience and little alternative but to follow the advice of their elders. The difficulties arise later for those restless few of any age whose vision is at variance with the custodians of the money supplies.

Most, if not all, funding agencies have far less money than they need to satisfy the demand for research support. Most, too, are publicly funded bodies that are required to be even-handed, so their usual response is to seek advice from the best established scientists on how the scarce resources should be allocated, and base their actions on consensus opinion. Most if not all private agencies behave similarly. These procedures yield the best advice on the next steps to take, and it is inevitable that most scientists will be strongly influenced by the collective opinions of their colleagues in their choice of problem.

'Next-step' or evolutionary science is, of course, very important. It secures the foundation of what we know, and steadily expands the possibilities of technology. We shall, however, have little to say about it mainly because next-step science usually means a choice between many options, when issues other than scientific often dominate: industrial, social, economic, etc. The goals are often clear—how to achieve room-temperature superconductors say, or to produce disease-resistant crops—even though they may require the very highest levels of intellectual commitment and ability to achieve them. The choice will rarely be a free one, however, and will usually be severely restricted by the type of equipment the scientist or small group already possess.

In 1972, when I was an elementary particle physicist working at the Daresbury Laboratory, my research group wanted to measure an exotic property (the axial vector form factor) of a nuclear particle by what we considered to be an ingenious route. It had to be ingenious because we didn't have the right type of spectrometer, and I wanted our group, and our laboratory, to steal a march on our rivals at the

Deutsches Elektronen Synchrotron—usually abbreviated to DESY, in Hamburg, who had the equipment but apparently had not yet had the idea. The idea was very much next-step although I did not think in those terms at that time. For me, the only things that mattered were that the measurement might turn out to be important, it had not yet been made, and that we might be the first to make it.

I went to see the Laboratory Director, Alick Ashmore, to get his support. I had expected that, as times were hard, it would not be easy to convince him that the work should be done, that we could do it, and that we should not be given the modest resources necessary to do so. I had also expected that I might need more than one bite at the cherry, for Alick Ashmore was often a hard nut to crack, though he was always fair. To my astonishment he was against the idea from the start, not for any of the reasons I had anticipated but because he thought that in the circumstances we should let DESY do it because their superior equipment would enable them to do a better job! I was flabbergasted because he had implied an argument I could not even contemplate: that is, experiments in science should be, in effect, put out to tender, and success should go to the best-equipped group. I could not budge him, although so far as I was concerned he might as well have tried to convince Robert Falcon Scott to abandon his bid for the South Pole in favour of Roald Amundsen because the Norwegian's equipment was better suited to the conditions.

We have come a long way since the days when every scientist was free to do as he or she pleased. There are now endless reports to write and committees to convince before approval for an idea can be obtained, if you are lucky. The problem is that for any position in the sciences, there is a very large number of possible next steps, each of which, in principle, might yield an interesting result. Research is also intensely competitive. There are more than a million scientists world-wide, and most of them struggle to find a piece of unoccupied scientific ground on which to focus their attention. The net result of all this activity is the provision of enormous densities of information in a few dozen well-trodden areas, while vast tracts go unexplored. It is there-fore inevitable in these circumstances that sooner or later the right to carry out experiments will be decided mainly on grounds of efficiency if we are to extract the best value from the now severely limited resources available.

In building a career, therefore, the first goals to be achieved are to convince your colleagues and your superiors that you are efficient, effective, reliable, and if possible that you might have flair, no matter what you choose to do. In these respects, and in many others too, building a career as a scientist is no different from any other profession. Somehow you have to get noticed, and you have to impress if you want to get on.

If you choose the academic sector, which includes not only the universities but also the many government-funded research institutes like the Daresbury Laboratory, the Cambridge Laboratory of Molecular Biology, or the National Institute of Standards and Technology in Maryland, and many others that are run largely along academic lines, your scientific career proper generally starts in earnest only after you have been awarded a doctorate or a master's degree.

When you have finally written your Ph.D. thesis (some universities call it a D.Phil.) you will have to endure the trauma of the oral examination. Some universities appoint external examiners specifically for each examination. They may be experts with an international reputation in your subject, or they may be scientists with a wide general experience, but their main responsibility in these matters is to ensure that the high standards of the doctorate are maintained. In Britain, you may have to face only one or two keepers of the gate to your future career, and the meeting will be in private, but in Europe and North America, there may be a larger number of examiners, and in addition, you will probably have to defend your thesis in public.

The private oral examination typically lasts one and a half hours, but there is no set time limit. The atmosphere the examiners create varies enormously. It may be an experience you will never forget as your examiners take you virtually line by line through your thesis; or it may be relaxed and gentle, but your examiners will want to be convinced that the work is indeed your own, that you really understand what you have written, and that you are at least on nodding terms with the main issues in your field. Your supervisor will probably also be present to hold your hand and to throw you a lifeline if you might be wallowing in the waves of academic assessment. Otherwise, supervisors may be little more than anxious spectators as they watch their protégés being put through their paces.

The most spectacular public examination I have attended was held in 1989 at the Eindhoven University of Technology, where my friend Netty van Gasteren took her Ph.D. In the Netherlands, incidentally, those who have been accepted on a course leading to a Ph.D. are given the *doctorandus*, usually abbreviated to Drs. Netty was the student of Professor Edsger W. Dijkstra whose work we discussed in the last chapter. The Ph.D. examination was held in the University's largest

auditorium, and at the appointed hour a solemn procession of twelve capped and gowned examiners from far and wide was slowly led into the hall by a black-stockinged, frock-coated official known as the Pedel who, once the examiners were seated, banged his staff on the floor as a signal for the proceedings to begin. Faced with an audience of two hundred or more, and before the eagle eyes of the inquisitors, Netty introduced her thesis. There then followed a series of probing questions, from each of the twelve examiners in turn, sometimes in English, but mainly in Dutch. Precisely one hour after the proceedings had begun, and while an examiner was still in mid-sentence, the Pedel reasserted his authority by again attacking the floor with his staff before leading the dozen distinguished dignitaries deliberately to their tea.

During your Ph.D. course you will have been very much the servant of your supervisor (some would say slave) and expected to work at least a hundred hours a week. For example, Sam Ting, the American particle physicist, drove himself and his assistants (who were not all students) very hard indeed. During the 1970s, at least, he insisted on knowing and approving what they were doing with their own time as well as 'his', and sometimes called group meetings at four in the morning just to keep them on their toes. He went on to win the Nobel Prize for Physics in 1976.

Ting is rather exceptional, but in general, research students are the unsung workhorses of the scientific enterprise. They are expected to work for little more than a subsistence allowance (in London in 1992 typically £4800 a year; in New York graduate students, as they are called there, earned about $12 000 a year) and yet their 'ragamuffin, barefoot, irreverence', in the words of Jacob Bronowski, makes a major contribution to keeping the scientific flames burning brightly. There is, however, little in the way of smouldering resentment. Quite the contrary, I have never met a research student who has been other than reasonably happy with his or her lot: any discontent among their ranks arises from what they see is happening to their elders, and therefore might happen to them in the future.

Humble though they may be, every research student has a chance of becoming a Cinderella and striking it rich. In 1965, Jocelyn Bell graduated in Natural Philosophy at the University of Glasgow, one of the few universities which still uses the old name for the sciences, although in this instance the subject is really physics, and later that year began a Ph.D. course at the University of Cambridge. She had decided to be a radio-astronomer, which at that time was a very young field. Most stars, including the Sun, galaxies, and other stellar objects emit radiation throughout the entire spectrum of wavelengths, from radio waves at the very long end to microscopic X-rays at the other, with the familiar visible radiation in between, and each type of wave yields information that is characteristic of the process which generated it.

Jocelyn Bell joined Professor Tony Hewish's group. He had recently decided to build a new telescope of simple and elegant design to study quasars, those very remote quasi-stellar objects whose awesome power in comparison with our own sun's feeble flame had been discovered only a few years earlier. The Cambridge radio-telescope was in principle a simple affair. It consisted mainly of several miles of copper wire strung across hundreds of poles a few yards apart arranged regularly in a four-acre field, and it relied on the Earth's rotation to scan the heavens with its gaze. Jocelyn Bell and half-a-dozen other students laboured for nearly two years to build the instrument (work which she must have felt was a far cry from the intense intellectual effort she had signed on for). Her experience, however, is quite typical of the so-called 'big sciences', astronomy and high-energy physics for example, which need very sophisticated equipment, often on a large-scale, that might be designed and built mainly by the team itself, with research students providing the bulk of the labour.

In 1967, the telescope was finished but it had to be tested and de-bugged, and Jocelyn Bell had to prove that the squiggles on the pen-recorder output were actually caused by radio waves from distant stellar objects rather than from, say, a nearby television station. This is by no means a trivial problem: the signals they were looking for are extremely weak. Noise irritates most people, but for scientists 'noise'—which is shorthand for unwanted signals—comes in many forms, and is one of the most formidable obstacles which stands between them and a credible and reproducible measurement. As anyone who has ever searched for four-leaf clovers will know, it seems as if almost every plant in the field is trying to persuade your eyes that it is exactly what you are looking for. Electronic noise is no less devious, and the world is bathed in it. The team therefore had to ensure that the tiny whisper of a stellar signal could be heard above the hectic hiss of the neighbourhood noise.

It was not long before 100 feet of paper chart was spewing from the telescope's pen recorder every day. The would-be Dr Bell was to have sole responsibility for analysing all this output, for a time at least, because it would provide the material for her thesis. Although she had just spent some two years slogging away in a muddy field, that would hardly count towards her doctorate. Her real work was to measure the apparent angular diameter of as many quasars as the telescope could see and she could analyse. Once she had collected the data she needed, the telescope would be passed on, so to speak, to the next student in the line.

The paper chart was analysed by hand rather than automatically by computer, which was a very wise decision because computers can at best only do precisely what they are told, and anything unexpected will usually be automatically ignored. The first seeds of suspicion were sown in Jocelyn Bell's mind when she came to the data from 6 August 1967 and noticed 'a bit of scruff' she could not account for. She had worked out that the blip next to it was caused by interference, but the 'scruff' was not quite the same shape. Perhaps it was caused by interference from a passing car; or chance reflections from a nearby corrugated iron roof; or perhaps it was from a satellite? But some six weeks later she had a similar sighting, and from its timing realized that the mysterious signal seemed always to be coming from the same part of the sky. Could it be a message from extra-terrestrial little green men? Perhaps she was getting carried away, and was being fooled by something terribly obvious and mundane. If so, bang would go her reputation for reliability and efficiency when her betters pointed it out.

She agonized over what she should do, and although many others might have swept the scruff under the carpet as yet another glitch, she decided she would tell Professor Tony Hewish about it. From a research student's point of view, supervisors are like ship's captains, friendly enough at times, but not to be bothered with trivia. Her heart must have been in her mouth as she showed him the 'scruff', but he took it seriously, and they decided to take new measurements with more sensitive equipment. However, the scruff was not to be seen. 'You've lost it,' Hewish complained, but she persevered until the weak signal at last managed to peep above the noise. There it was pulsing every one-and-a-third seconds as regular as clockwork (Fig. 5).

It was then her supervisor's turn to get cold feet. What could it be? Might it really be little green men? What would become of his considerable international reputation if they were to publish before they really knew what was causing it? Perhaps the scruff was caused by interference from another radio-astronomer's equipment? He got in touch with his friends at the other labs in the UK to ask whether they had been doing anything that might cause it, and perhaps also to gauge whether or not they had seen anything strange, but he drew blanks everywhere. She made further measurements, and discovered what seemed to be another three sources of scruff. Every check had been

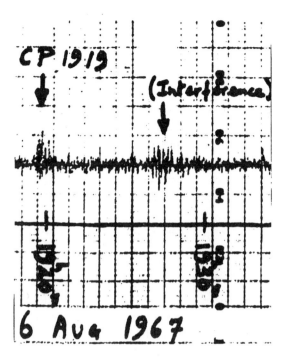

Fig. 5. Signals (CP 1919) from the first pulsar to be discovered. The illustration shows a couple of inches of chart recording from the miles of recording that Jocelyn Bell had to examine for her thesis. Courtesy of the Cavendish Laboratory, Cambridge.

made, and so they drew deep breaths and in February 1968 published in the journal *Nature* news of their pulsating radio stars, later called pulsars, which soon flashed around the world.

Explanation of their remarkable phenomenon was soon forthcoming. Only a few months later Thomas Gold of Cornell University suggested that Jocelyn Bell's 'scruff' came from a distant rotating neutron star. When stars die they may take any one of a few exit routes. They may for example slowly cool down and fade away, or more flamboyantly they may explode as a supernova, leaving behind an enormously super-dense mass of neutrons closely packed together in a shrunken space of smaller size than say London. As the star contracts, like an enormous twirling ballerina pulling in her arms, its rotation rate increases, and the expelled electrons are held increasingly tightly in the grip of a super-intense magnetic field. As Heinrich Hertz had discovered in 1888, rapidly moving electric charges emit radio waves, but since, as on Earth, the magnetic pole of the pulsar would not coincide with the geographic pole around which it rotates, we would 'see' the radio waves from the orbiting electrons only when the star's magnetic axis flashes through our lines of sight, like some stellar lighthouse. Today, some hundreds of pulsars have been discovered in our galaxy, and a few rotate hundreds of times faster than the ones that Jocelyn Bell had found.

The discovery of pulsars gave the first confirmation that neutron stars really do exist, and Tony Hewish won the Nobel Prize for Physics in 1974 in recognition of his seminal work. The award was not only for the discovery of pulsars. The Physics prize that year was shared with Sir Martin Ryle, who earlier had pioneered radio-astronomy, and together they had helped to lay the foundations of this important new field.

Not surprisingly the Nobel Committee came in for a great deal of criticism for not recognizing Jocelyn Bell's contribution. After all, had she not shown great perseverance and courage the discovery of pulsars would probably not have been made at that time. Indeed, it seems that scientists at Jodrell Bank had seen a similar signal some years before but had discarded it as unimportant. She was, of course, only a research student, but the Prize had been awarded only the year before to Brian Josephson for work he had done as a research student, also at Cambridge, on superconductivity.

Quality cannot, however, be reliably measured, and it is one of the disadvantages of all awards that the chosen few are often separated from the possible many by only the finest of dividing-lines, which might be drawn differently on another day. It is a pity that Jocelyn Bell slipped across whatever lines were drawn in her case, not least because she is a woman. Had she won a share in the Prize it might have inspired many more young women to compete in the predominantly male scientific world of the physical sciences. However, Jocelyn Bell (Professor Bell Burnell as she now is) herself shows not a trace of rancour. Prizes do not really matter. Nothing can alter the fact that she was the first earthling to pick up a signal from a neutron star and do something about it.

Careers are rarely established in such style or so quickly. Normally, it takes several years of hard grind to build a reputation which, as for other professions, is a scientist's most treasured possession. As an academic scientist, your first professional appointment will usually be a post-doctoral fellowship of some kind that you may have seen advertised, or heard about through your supervisor's connections. There will be the usual interview, and you will need to shine. Most post-doctoral jobs are for two or three years, and since each one can take up to a year to locate and win, post-docs are almost continually considering their positions.

The uncertainty and frequent movement are unsettling not only for the young. Many research projects, especially if they are ambitious, will take longer to bring to fruition than the time-span of undivided attention that post-docs can usually give, and research can easily stall when an itinerant post-doc moves in pursuit of what seem to be better prospects elsewhere. The frustrated supervisor then has to cast about for suitably qualified replacements, who in their turn will need some months before they can settle in and contribute effectively to their new work, and who, of course, before long will also start keeping a look-out for the next stepping-stones in their own careers.

In spite of all this trauma, the passion and excitement of science usually remains undiminished for post-doc and supervisor alike in their quest for new scientific territory 'which is so virgin that no eyeball has ever set foot in it', in the words of a Yale research student. Everyone is aware of these inefficiencies and does their best to cope. Professor Peter Riley sent me a lovely notice from the University of Texas at Austin: 'The floggings will continue until morale improves.'

A few decades ago, the progression from post-doc to a tenured appointment such as a lecturer or an associate professor in a university, or the equivalent government-funded institution, was almost automatic if you were good enough and were prepared to move about. Since then, although we have seen unprecedented growth in terms of the number of people employed and scientific output, the number of universities and institutions has hardly changed, and one of the results of increasing budgetary pressures is that more post-docs are chasing fewer permanent positions as posts are frozen or abolished. Post-doctoral salaries are usually related to age, and when money is tight, older more experienced hands can be an expensive luxury for hard-pressed group leaders. In consequence, as each birthday passes post-docs become increasingly concerned that they might be priced out of the market; and as their anxiety grows they are tempted to cast their nets ever wider in their search for so-called security of tenure.

One of the keys to promotion is your publication record. In general, the more papers you publish the better your chances will be. Unfortunately, this is not always consistent with good science; papers published without long and mature reflection are usually quickly forgotten. There are exceptions, of course. Albert Einstein spent his first post-doctoral years as Technical Expert (Third Class) at the Berne Patent

Office. He had been awarded his Ph.D. by the University of Zürich, but, in the words of his biographer Ronald Clark, that university, among others, saw him as 'an awkward, slightly lazy, and certainly intractable young man who thought he knew more than his elders and betters' and so passed him over.

However in 1905, when he was 26, Einstein published three papers in a single volume of *Annalen der Physik*, any one of which would have made his name. The first was on what is now called the photo-electric effect, and eventually led to his Nobel Prize in Physics in 1921; the second was on the Brownian motion, and showed how the collisions of individual molecules can be observed in liquids; the third was on special relativity, and its revolutionary vision shot him into fame and bitter controversy.

It is difficult to overestimate the importance of Einstein's work on the special theory of relativity and, over the following decade or so, on the more widely applicable general theory which includes gravity. In classical physics, space was thought to be absolute and universal—it was called 'the firmament' for many centuries—and time was thought to flow uniformly through all of it like a vast tide. Einstein revealed, however, that the four dimensions of space and time are inextricably interrelated and form a fabric called spacetime that can deform when matter is present in much the same way as a person's weight depresses a cushion. Thus, the quantity and proximity of the matter present in a particular place determine the curvature of spacetime and the apparent rate of the passage of local time as measured by a remote observer. Furthermore, the all-prevailing and totalitarian realm of time was now shown to be somewhat diminished in stature, with each observer's local colony being 'governed' by its own time lord. The differences from the classical picture are detectable only with sophisticated equipment, but Einstein's work, as did Newton's before him, enabled the first steps to be taken towards embracing and unifying what hitherto had been disparate phenomena; in Einstein's case those of electromagnetism and gravitation, although the journey is still far from complete.

The differences are small because the stiffness of the fabric of spacetime is largely determined by the magnitude of the velocity of light, which in the classical view of space and time has no special role or significance, and in some circumstances is implicitly required to be infinite. This requirement arises because if time progressed at the same rate everywhere, there would have to be a way of verifying this would-be state of timely conformity because Nature is never secretive. If, however, we were to turn a powerful telescope on to the Andromeda Galaxy or some other remote object and saw at a stroke that clocks there were in time with clocks on Earth, communication would have been achieved instantaneously. But it had been known for over two hundred years before Einstein came along that light does not travel with infinite speed, and so Einstein's youthful genius was all the more remarkable. Thus, Nature has no way of always keeping everything in the universe in timely step, and consequently we all march to our own relative time.

The velocity of light is very large (300 million metres per second), and so relativistic effects are not obvious in everyday life. Nevertheless, every molecule of our bodies, and everything around us, is embedded in the fabric of spacetime, and we cannot escape whatever effects its structure might cause, even though we might not be aware of them. Einstein's ideas, however, were so radical, the classical thinking so engrained, and the mathematical problems of dealing with a four-dimensional universe so difficult, that even today there are few areas of science where the new thinking has been fully accommodated. Quantum mechanics, for example, on which most of our understanding of atoms and molecules is based, still generally treats time as if it were an optional extra and space as if it were not affected by gravity. Who knows, therefore, what new insights will come when we eventually learn how to adjust, and what their consequences might be?

The relativistic controversy went on for many years. Ronald Clark, describing Einstein's somewhat lukewarm reception during a visit to Paris in 1922, commented 'He was the man who had upset the scientific applecart—or at least appeared to have done so—and he naturally aroused the resentment of those who believed in things as they are. He was not only a scientific iconoclast but German as well. To compound the crime, he was not only a German but a German Jew.' These remarks can perhaps be fully understood only by recalling that Einstein's visit was made only four years after the ending of one of the bloodiest wars of attrition in history. It has always been difficult to separate science from considerations of politics and religion, a subject to which we will return in Chapter 8.

Einstein's scientific achievements were hardly average, and yet post-docs today are expected to publish almost as frequently as he did in his miracle year, and at least to maintain a high standard if not to win awards. University departments like biology or chemistry cover an enormous range of intellectual ground; those who study, say bacteria, or fungi, or plants in general, or animals, or molecular biology, may all belong to the same department but may have little expert knowledge in each others' fields, or have anything more than a nodding acquaintance with the changing stream of itinerant post-docs associated with them. When senior staff are considering candidates for promotion, therefore, a lengthy list of publications may help to tip the balance when there are arguments over whose protégé should win recognition, or when otherwise there are disagreements over which candidate should be given a job.

This is not as frivolous as it might sound: 'publications' in this context is a shorthand for papers published, or accepted for publication in a refereed journal. Publications record your priority, and reasonable steps are taken to ensure that the research you wish to place on record is indeed a scoop, so to speak, and that no other journal elsewhere has printed the same or similar results.

Your paper will have to be prepared in the style of the chosen journal, and should fully acknowledge previous work that might have provided the launching pad for your own efforts, a duty that will be a delight if the earlier work just happens to be your own. It should also be clear and understandable, though the great majority of papers are virtually incomprehensible to other than the *cognoscenti* of the field.

These and other guidelines are overseen by independent referees, and your paper might be sent to two or three of them for their views on its suitability for publication. Independence here refers to referees' relationships with the journal; for your papers will almost certainly be refereed by your competitors. Their opinions are taken very seriously by editors, and your paper might not be accepted if they believe that its claims are unsubstantiated, or they are trivial, or otherwise they do not like it. Their opinions may be passed on to you, but hitherto they have usually been anonymous, a practice which seems to be changing. Although it is not impossible for you to challenge their conclusions it is certainly not easy.

The very prestigious and international journals, like *Nature* and *Science* and a few others in each of the main disciplines, set an exceptionally high standard, but otherwise outright rejection is rare; referees will usually make a number of suggestions on how a paper might be improved, and in general try to be helpful. Referees, who are usually drawn from the ranks of established scientists, take their responsibilities very seriously, and may devote several days to studying a manuscript, even though they will not be paid for this labour of love. But there can be serious problems, especially when a paper is highly controversial or perhaps might be a breakthrough. In this case you may have to hawk your paper around. Opinion varies among scientists generally about whether referees' comments should be attributable. There is little doubt that a tiny minority take advantage of their anonymity to settle old scores, and that some quite simply fail to understand what they have read, but by and large the system works well.

The number of journals from which to choose is truly enormous. Ann Okerson from the Association of Research Laboratories in Washington D.C. estimated in 1992 that the world total might be some 250 000. The fragmentation of the disciplines has arisen largely along arbitrary lines, and there are journals that not only cater for the main divisions but also seek to embrace much more broadly and arbitrarily defined subject areas such as cell biology, or the chemistry of new materials. New titles appear almost daily, especially in the biological sciences, covering such areas as 'DNA sequencing and mapping', or 'computer simulation of biomolecular systems'. The profusion of new journals perhaps reflects both great activity and uncertainty in these fields.

Most English-language journals seek an international coverage, but in addition there is a vast literature in other languages which is usually more geographically specialized. Many of these journals hardly circulate outside their native countries, and it is almost impossible to keep oneself abreast of them.

Not surprisingly, not all journals bestow the same stamp of approval on your work. Many scientists are loyal to an informal caste system among the journals, and would not dream of allowing their work to grace the pages of other than the most rigorous and prestigious of them; furthermore they would tend to judge others by the journalistic company they keep.

Regardless of the journal, it seems almost inevitable nowadays that when your publication finally appears in print it must be phrased in terse, turgid, and technical prose, with little to indicate to the uninitiated that you have discovered a facet of Nature's behaviour which no one else has ever seen. 'Gee-whiz' journalism would be equally out of place, but such sterile styles can hardly help to extend science's readership. News of discoveries is usually reported in such a way that only the specialist can detect whether the result being celebrated is a *tour de force* or a triviality. Neither will there be anything to indicate the difficulties you might have had: the false starts, the mistakes, the problems with reproducibility, the failure of equipment, and the hundred and one other things that will have gone wrong. Rather your paper will record an uneventful journey as if your Everest had been conquered by way of a carpeted staircase to the summit. It was not always that way, and perhaps it may change again in the future.

As a post-doc you will normally be given responsibility for writing at least the first draft of your group's prospective publications. Your name may appear in alphabetical order; or it may be first, depending on your group's protocol in these matters or the significance of your contribution.

All this writing will stand you in good stead when you are lucky enough after a few astute job moves to win a tenured position. Before that, responsibility for choosing a problem and raising the money to work on it will usually have been someone else's. A tenured position does not necessarily mean that you have to set up on your own. You may decide to continue as you were, especially if you have been working with someone inspirational or famous. Some collaborations can be for life if the personal chemistry is right; and, conversely, their breaking up can be just as devastating as divorce.

Fund-raising will probably be the most difficult professional problem you will ever have to face on a regular basis, and yet you will be fortunate indeed if you have had any training for it. Like much else in life, it is a talent you are supposed to be born with. However, once you have selected the problem you want to attack, your next task will be to prepare a proposal for the agency you believe might support the work.

In Britain and some other countries, research funding tends to be monolithic. The agency will usually be one of the government-sponsored Research Councils or its equivalent, although there are some industrial and other private agencies that can help. In the United States, apart from the large agencies like the National Science Foundation, the Department of Energy, and the National Institutes of Health, there is a profusion of private and industrial sources of finance, and scientists in some fields find it easier to persuade a number of agencies to provide a portion of the cost of a programme than to win the total from one. This means a lot of persuasive writing, an understanding of the objectives of each target organization, and a fine sense of timing, but once the first pickle has been sprung from the jar it is often easier to persuade others to follow suit. All this requires skills beyond those traditionally associated with a successful scientist; it has indeed been said that no other scientific community depends as much on professional entrepreneurship as does that of the United States.

For the big sciences it is a different story. Almost all the very large amounts of money required come from government sources. If the programme is specially large, collaboration between governments might be required. The European Nuclear Research Laboratory founded during 1953–4 in Geneva, and usually referred to as CERN, had in 1992 a budget of some £350 million shared among 17 member states. It is perhaps the most successful example of international scientific collaboration, and was set up '. . . to study phenomena involving high energy particles in order to increase the knowledge of such phenomena and thereby to contribute to progress and to the improvement of the living conditions of mankind . . .'.

CERN's crowning glory was the discovery in 1983 of the W and Z particles by Carlo Rubbia and his group, for which he won a share in the Nobel Prize for Physics in 1984. The discovery is important because it provided clear evidence of the long-suspected unity of character of three of Nature's forces: the strong, the electromagnetic, and the weak. The fourth force, gravity, remains stubbornly outside this framework for the time being. The discovery was made by using colliding beams of protons and anti-protons, and was a *tour de force* not only for Rubbia, but also for Simon Van der Meere and his team who designed the highly efficient anti-proton accelerator that made the experiment possible. As a result, Van der Meere shared the Prize with Rubbia.

Big science not surprisingly therefore is highly political, and usually only the most senior and influential scientists participate in decisions on which major projects might be funded or discussed with other governments. Many of the scientists who will actually do the work, and whose careers are often on the line, can usually do little more than watch and wait.

The sums involved can be gigantic. The United States is building a Space Station which is estimated to cost $37 000 million. A decision has also been taken to build a Superconducting Super Collider at Waxahachie in Texas at a cost of more than $8000 million. A condition of the approval of what will be the largest particle accelerator in the world was that at least a third of the cost should come from sources other than the US Federal Government (see note p. 92). There are very few countries rich and far-sighted enough to consider such high stakes, but in any case most scientists can hardly expect to have much influence on the diplomatic and political horse-trading that inevitably will be needed if any negotiations are to be successful. The high-energy physicists are a possible exception in that they are renowned for their powerful lobbying techniques. It was said in the 1980s that the USA faced three superpowers: the Soviet Union, China, and the high-energy physics community. Major projects are not confined nowadays to the physical sciences. The U.S. Congress approved in 1990 the Human Genome Project, a $3000 million effort over 15 years to map and sequence the DNA in an entire human genome. Other nations have approved programmes of similar relative size.

Participation in major programmes like these can be very exciting and stimulating. They are in many ways the scientific equivalent of a trip to the Moon in terms of their magnitude and complexity, but the numbers of people involved can be large, and one's sense of individuality and direct personal association with the problems being tackled tend to be lost. Major programmes also require a sustained commitment over many years. Rubbia's group, for example, included 200 physicists, and the discovery of the W and Z particles was an experiment almost 10 years in the planning and execution, and in the analysis of the voluminous data. As this is a good slice from a scientist's active and productive life, it is not surprising that one finds the highest levels of professionalism and dedication among the people involved, whatever the personal costs, and the Ting approach, mentioned earlier in this chapter, becomes understandable. A casual visit at any time of almost any day or night to one of the major centres of big science will usually reveal a hive of activity, as everyone concerned does their utmost to ensure that their own major investment pays off in some way; in science, as in other walks of life, there are definitely no prizes for coming second.

The vast majority of scientists from other disciplines have much more control over their destinies, at least in principle. This caveat arises because although scientists are free to decide what they want to do, it is the funding agencies that actually decide whether they should be allowed to do it. A key element in their decision-making is the process called peer review. It is a process familiar to everyone from the hardworking cook to the most powerful presidents and prime ministers. We are all conscious of the opinions of those around us, whether it is on the taste of our food, or the clear-headed charisma of our leadership. Politicians are continually preoccupied with public perceptions of their performance, especially at election time. They are professional communicators selected for their persuasive abilities among other things, and nowadays they have the services of marketing people and media specialists renowned for their genius in positive thinking in adverse circumstances. Scientists, on the other hand, are ultimately and solely answerable, so to speak, only to Nature, who will not be swayed by any power of persuasion, no matter who you are or how many people hold a similar point of view.

In 1968, I attended an international school of physics at Erice in Sicily. It had been organized by the ebullient and energetic Antonino Zichichi, who has created one of the most important and wide-ranging scientific schools in the world. It no doubt owes much of its success to his astute choice of location, which is a charming medieval town overlooking Trapani and the blue Mediterranean.

One of the school's tutors was Tsung Dao Lee, who in 1957 had won the Nobel Prize for Physics together with Chen Ning Yang (who likes to be known as Frank in honour of Benjamin Franklin) for their work, the previous year, that pointed out on theoretical grounds that a quality called 'parity' might not be conserved in interactions involving the weak force. The concept of parity is normally important only for matter on an atomic scale and is related to the differences in behaviour seen by an observer who looks at a system either directly or indirectly using a mirror. In the everyday world, dentists usually assume that a mirror image of what they are doing to your mouth is a perfectly accurate representation of the real thing. Thus, parity might be said to be conserved in dental interactions. But Lee and Yang were claiming that for some types of radioactive decay the mirror image of these weak processes would not be allowed by Nature. The idea that Nature is not always even-handed with respect to left or right, or clockwise to anticlockwise, was astonishing, but it was confirmed experimentally by Madame Chieng Shiung Wu, also in 1957.

During the Erice school, Lee posed a subtle and elegant question on time-reversal invariance that need not concern us here, except to say that it had his audience stumped. The arrow of time has an inexorable quality, as we all know, but for elementary particles the situation is both more open and complex. We were invited to present our solutions while Lee sat in faintly amused silence as we picked out our clumsy tunes at the feet of the master. A wide range of solutions was suggested and we could not agree. What was to be done to resolve the divergence of our views? 'Let's take a vote,' someone joked, at which the normally serene Lee immediately snapped, 'Science is not a democracy.' He went on to tell us that voting would be futile on this occasion, for we were all on the wrong track, and he proceeded in his characteristically lucid style to show us the error of our ways.

Lee's succinct remark about science is applicable to many facets of life when action has to be taken in the face of controversy. In economics, for example, there have been many theorists, but few theories have stood the test of time. After all, the world is highly complex, and we understand very little. Democracy as we know it works best when decisions have to be taken on such issues as taxation, and can hardly be expected to shed new light on old problems.

Individual scientists have no doubts about Nature's indifference to popular opinion, no matter how well informed. But the scientific enterprise today is controlled not by individuals but by committees, these relatively modern institutions which, in the words of Sir Barnett Cocks, a former Clerk of the British House of Commons, are cul-de-sacs down which ideas are lured and then quietly strangled.

The largest granting agencies may have many dozens of committees to decide which research proposals shall be funded, and how much money they shall each be allocated. Committees are assembled from the ranks of practising scientists in the discipline or sub-discipline over which they preside. Members are normally appointed in the first instance for periods of one to three years, and appointments may be, and often are, renewed. Scientists often complain of the additional workload that committees impose, but they are nevertheless usually eager to serve, even though the work is generally unpaid.

This type of peer review is by far the most important and influential of the forms in operation today. For publications, we have discussed some of the ways in which peer review operates. Referees may require changes to manuscripts, and in some cases may even reject them, but there will almost always be a journal willing to publish, and no-one can alter the fact that the work has already been done, whatever the reviewers might think.

For research *proposals*, however, one or more peer review committees, depending on the size of the grant being requested, must be convinced that the proposed work is worth funding *before* it can be started, a process that would be more accurately called peer *preview*. Proposals are submitted in writing, and decisions are usually based on the consensus opinion of the committee members based on their own deliberations, and also on reports from referees. Committees meet typically once or twice a year, and for the larger agencies, each committee may have fifty or more proposals to consider. The proposals themselves are normally substantial documents, and there are usually several referees' reports for each one of them. Committee members are among the most reputable of scientists, and hardly have time on their hands, yet the peer preview system is based on the assumption that they will have assimilated their ponderous portfolios before meetings take place. It is a formidable tax on human frailty, but even though scientists are often superhuman in their dedication, and will burn vast quantities of midnight oil to meet their deadlines, there is little they can do to increase the number of hours during which a committee will meet.

Consequently, the time allocated for the discussion of some proposals can be trivially short.

Research budgets nowadays are very much less than those required to fund every proposal that would survive reasonable standards of scrutiny, and committees are frequently forced to create criteria with hyperfine (some would say fictitious) gradations in quality to separate the outstanding that can be funded from the merely excellent that will be rejected. Sometimes the slightest shadow of a doubt expressed by a committee member is welcomed with relief by the rest, since it will be sufficient in these circumstances to cast one more otherwise good proposal into oblivion. The situation is no less serious in the United States. At the annual meeting of the American Association for the Advancement of Science (AAAS) held in Washington in February 1991, it was pointed out that almost 100 000 people were involved in writing peer-review assessments of proposals that were not funded. It was suggested that much of this wasted time could be saved by assessing proposals only for their technical feasibility, and putting every proposal passing this test into a lottery. Proposals coming out of the hat before the available money was spent would be funded. The AAAS has no executive responsibility for grants, but these perhaps tongue-in-cheek observations made by senior scientists are indicative of the malaise.

Nevertheless, in building your career, you will have to become adept at writing successful proposals if you are to progress. At a meeting held at a very presitigious American university in 1991, the Dean of Graduate Studies told a meeting of junior faculty 'Forget Nobel Prizes, forget awards of any kind. Your job is to find out what is being funded, and go out there and get some of it.' On another occasion at the same university, a Vice-Chancellor overheard a senior faculty member excusing himself from a meeting to get back to the laboratory. The Vice-Chancellor said, 'Full professors should not spend any time in the lab. That's what post-docs and graduate students are for. Your job is to get funded, and that together with your teaching is a full-time job.'

Even at the best of times, peer preview encourages scientists to be cautious, and to ensure that their proposals will fit nicely within the disciplinary boundaries of the target committee. As we have discussed, the disciplines and their many subdivisions are themselves frozen snapshots in time of current misunderstanding, and so the more progressive and inspirational your ideas might be, the more likely they are to be excluded from the frame. Louis Alverez, the American scientist who won the Nobel Prize for Physics in 1968, and who was also famed for his wide-ranging interests, goes much further. In his book *The adventures of a physicist* he says that, 'In my considered opinion the peer review system, in which proposals rather than proposers are reviewed, is the greatest disaster to be visited upon the scientific community in this century.'

Your 'track record' will also be taken into account, which means that if you are an older researcher, it will not be easy to change your field. For young people, there are arrangements that provide start-up awards, but they are usually modest, and are intended to help you to get established. It is virtually impossible, however, for a young person acting independently to get backing for a substantial research project from the traditional sources of funds, even though it is well known that the young can be highly creative and in the past have made many of the major and revolutionary discoveries that have transformed understanding. There are many private organizations and foundations that make special provision for gifted scientists, young or otherwise, particularly in the United States. Selection is almost invariably by peers. These bodies are disciplinary or geographically orientated, and sometimes their deliberations are in secret. Your award may therefore arrive out of the blue, like manna from heaven, but you will need to be famous or well connected to win it.

Even if you do, an intrepid scientist cannot easily escape the tentacles of tradition, whatever the sources of funds. In 1982, I met a young assistant professor at one of the finest universities in the USA. His proposal was exciting and ambitious, although it took several discussions before I saw the point. But just as we were about to agree to fund it, he withdrew the application, saying that he had been advised by his faculty colleagues that the work could damage his career if it did not work out. It would be best, he said, if he concentrated for the time being on building a solid reputation as he could always come back to us later with his idea. He has not yet done so.

Conscience does make cowards of most of us at least, and we in Venture Research are always careful to ensure that rebellious young people in particular are fully aware of the risks they would take by challenging convention. In 1984 one of my colleagues, Keith Cowey, met Dr Steve Davies, a young organic chemist, during a routine visit to the University of Oxford, and we subsequently invited him to our offices in London. One of the first things he said was that much of the thinking in organometallic chemistry was wrong. This is the sort of remark we love to hear, especially if the disdain can be defended, which he proceeded to do.

Organic chemistry—the study of the almost limitless number of molecular compounds containing carbon—is one of the keys to life.

Nature, as we have discussed, is not even-handed. It has also been remarked that we need a three-dimensional picture rather than the fashionable, flat two-dimensional representations if we are to understand molecular and other natural behaviour. In the everyday world, objects like a ball, or a blade of grass, or some unsymmetrical things like a cup or a teapot are indistinguishable from their reflections in a mirror. But the mirror-world can be much more complicated. Two friends looking in a mirror can readily see each other's strange appearance, and while one's own face looks perfectly normal as one has nothing to compare it with, the reflection might seem to follow with its left hand everything one might do with the right. If that strange copy of oneself emerged from the mirror and became real one might not be too surprised if its odd behaviour did not match its appearance.

Molecular 'handedness', usually referred to as chirality (from the Greek word for hand), can be highly significant. Limonene, for example, is a molecule that is either left or right handed, but both forms otherwise have identical chemical formulae. Oranges and lemons each have an unmistakable and instantly recognizable aroma, but the molecular messengers that carry their perfume are limonene molecules of either one hand or the other. Nature has arranged that each fruit makes one type of limonene only, and to ensure that this supremely subtle chemistry does not go unnoticed, Nature has also equipped our nostrils with equally exquisite and efficient organs to detect each fruity fragrance and signal the news to our brains.

The sophisticated elegance of all this is not an isolated example of Nature's flair, but part of the normal service. Many biologically active molecules—proteins, carbohydrates, sugars, etc.—have a characteristic handedness that affects their performance. When the wrong-handed molecule for a particular process finds its way into a living organism, it will most likely be disposed of as waste. But sometimes the wrong hand can cause severe difficulties as, for example, in the case of Thalidomide, a pharmaceutical product marketed during the 1960s to alleviate morning sickness during the first few months of pregnancy. Unfortunately, the mirror-image version of the Thalidomide molecule we now know to be safe and effective was found to increase the incidence of malformations. The drug was withdrawn when its tragic side-effects became known, but for a decade or so there followed bitter controversy and litigation over such questions as who knew what, when they knew it, and who was responsible.

The sensitive dependence of molecular behaviour on chirality, or handedness, was not recognized during the 1950s when the drug was being developed, and the industrial processes used in its manufacture produced precisely equal quantities of left- and right-handed molecules as if they were gloves. But a glove factory could easily switch production if there were a sudden surge in demand from one-handed people, or for baseball gloves, for example. For chemicals, the differences between chiral species were far too subtle to be resolved by the production techniques of the 1960s, and even today such stereo-selective chemistry is generally inefficient or uneconomic.

Nature's methods are very efficient, but many of the processes of molecular recognition and chiral control are not understood. The specific enzymes on which these methods are based are the subject of intense world-wide study, not only in the academic sector but also in industry, and substantial progress has been made in understanding the general features of their structure and function. But enzymes are the ultimate in molecular specialists, and Nature uses a different type of enzyme for almost every chemical process she needs. Enzymes are also among the most complex molecules known, and it is unlikely that a direct and frontal attack on a problem of such vast complexity and variety will yield substantial improvements in the manufacture of pharmaceuticals in general, although there have been a few specific successes: the synthesis of L-dopa, for example, which is used in the treatment of Parkinson's disease.

Steve Davies told us that he had come across an apparently very simple solution to this enormous problem by tackling it from first principles. He had asked himself the question of how Nature might accomplish chiral synthesis if she had all the chemicals easily to hand that he had in his laboratory, rather than having to make do, as Nature must, with whatever materials can be found dispersed in the natural environment. He seemed to have discovered that a simple organic molecule with iron at its core behaved like a universal enzyme. It worked very efficiently, but he did not understand why it did so. We were astounded, not only by the audacity of his thinking, but also because the conventional funding agencies had declined to back the substantial and exciting programme on which he now wished to embark. Reasons for rejection are rarely given in such cases, but it is likely that his youth played a big part; young people are rarely authorized to attack major problems nowadays. Also, he wished to concentrate on understanding the processes of molecular architecture that lead to chiral control independently of where they might lead him, rather than, as is the current fashion in organic chemistry, to promise to produce a new product virtually every week. He was fully aware of the magnitude of this task, and the risks he was taking, but just to be safe, had decided on the prudent course of not neglecting his bread-and-butter chemistry while pursuing the *cordon bleu* prizes of his Venture Research.

We were delighted to be associated with him. Not only did his basic research promise to be of the highest quality, but the industrial possibilities it might create seemed extensive. Pharmaceutical processes that produce 50:50 mixtures of useful and useless products, as they generally do now, are wasteful at best, and efficient ways of controlling chirality could not only transform that industry, but might open up applications in other fields too. Many insect pests, for example, can breed only when the male picks up the scent of the appropriate female and follows it until he finds her. Scents of this type are called pheromones, and they are not only highly specific to each insect species, but they are chiral. The atmosphere is, of course, full of scented molecules of all types, not all of them pleasant, and so the insects have a signal-to-noise problem which they must solve if they are to survive. Nature has arranged that the male insect will be goaded into amorous action only if he picks up the right molecule *and* it has the right chirality. The slightest trace of the right molecule's mirror twin will cause the insect to smell a rat, and to run away. Some companies have tried to make insect traps using synthetic pheromones, but they cannot make them pure enough to fool the insects. Davies' methods should be much more effective, and if they were, would be a much better alternative to blanketing the environment with conventional pesticides. In other fields, materials made of molecules of the same chirality might have exotic bulk or surface properties; a new type of chiral switch might also be possible, which might lead to new ways to handle information; and there is always the unexpected.

The unexpected is indeed the outcome that many scientists crave, and for many of them the experiment or the theory that eventually results in merely adding another precisely polished stone to the foundations of what we already know contains an element of disappointment. Scientists are not unique in this respect. Travellers jetting around the world have long become resigned to the fact that the freshness, surprise, and wonder of new and colourful cultures that not so long ago they might have expected to enjoy without much effort can now be found only well away from the well-beaten tracks of tourism and business. For careers too, the differences between the major professional options seem to be disappearing. Not so long ago, one of the attractions of an academic career was the freedom it allowed that more than offset the relatively low pay, and in some countries, the low status too. That freedom is now under severe threat, and our societies are in danger of losing the precious asset of a sector that hitherto could be relied upon critically and rigorously to analyse and comment without fear or favour.

Now, in almost every country, freedom is increasingly being equated with the magnitude of the money supply. There is no doubt that in Britain and some other advanced countries, money for academic research and teaching is relatively very short. In these countries, there seems to be little awareness of the intimate links between science, technology, and economic prosperity, or of the other benefits to our culture and intellectual well-being that education brings. But beyond a certain rather low threshold, money does not necessarily convey freedom, nor the ability to live our lives to the full. There seems to be little diversity of life-styles among the very rich, for example. Money cannot create or satisfy Einstein's hunger of the soul; it must be allowed to find its own nourishment. In the following chapters we shall be discussing how best that essential condition might be met.

Note added in proof. The Superconducting Super Collider project was terminated in October 1993. The Space Station was still extant at that time, but the project was precariously balanced, having survived a passage through Congress by only one vote earlier that year. Its future probably depends on the extent to which international collaboration can be secured.

6

Industry: 'japanity' and the road to market

Industry has been one of man's pressing preoccupations since Adam delved and Eve span. It was industry that led to man's supremacy as a hunter–gatherer on land and sea; to agriculture and the first seeds of civilization; to the weapons that enabled the strong and ruthless to forge tribes into nations, and nations into empires. Through the ups and downs of history, industry's flame has flared and flickered, but despite its sometimes prolonged exposure to the bitter winds of tyranny and repression of virtually every description, its fire has never been extinguished.

All this ardour stems from man's refusal to accept Nature and the materials she supplies as he finds them. Until relatively recently, industry was based almost entirely on the inspired intuition of individuals, on their trials and errors, and on their dauntless determination to succeed where others have failed. The industrial revolution was powered by such single-minded visionaries, by their ability to inspire those around them, and above all by their skills in attracting the patronage necessary to make the jump from local curiosity to global celebrity.

Their genius was often drawn from prolonged practical experience in such long-established occupations as farming or weaving, but their frequent inability to describe what they had done in formal or theoretical terms did not mean that their discoveries were accidental or did not come from deep insight. Even though many technological pioneers were illiterate, their basic difficulty was that the conceptual frameworks embracing their work had not yet been developed, and their literary abilities were therefore irrelevant. Sir Christopher Wren, for example, was one of the supreme scholars of the seventeenth century, and together with others, helped to found the Royal Society of London. Shortly after the Great Fire of 1666 had razed the City of London to the ground, he became the King's Surveyor of Works, and supervised much of the rebuilding of the City's many churches. He also was the designing and supervising architect for St. Paul's Cathedral throughout the 35 years of its construction, and its dome in particular is an engineering masterpiece. These achievements, and those of every other architect before him, were made without explicit knowledge of the engineering concepts of stress and strain. They were not defined until more than a century later by Augustin Cauchy, a former engineer from Napoleon's army.

At the other end of the literary scale, George Stephenson could not read until he was 19, when he managed to find threepence a week from his engineman's wages to pay for night-school lessons. Nevertheless, he was shortly afterwards designing and building the steam engines that would eventually revolutionize transport the world over. Twenty years later (1824), this and other pioneering work inspired another French military engineer, Sadi Carnot, to begin to understand the rules by which heat sources can be manipulated, and to develop the science of thermodynamics.

Carnot's work, however, was ignored for many years. New theoretical insight always takes time to take root, but even Stephenson's tangible and demonstrable achievements were denied by the engineering establishment. They vigorously refused to believe, for example, that a steel-wheeled engine rolling on smooth steel rails would 'bite' sufficiently to create useful traction, though had any of them taken the trip to the Killingworth Colliery Railway in the north-east of England in 1816 they could have seen heavy coal wagons being pulled daily by Stephenson's locomotives at 6 m.p.h., notwithstanding the establishment's objections. Indeed, showing great foresight, Stephenson took trouble to make the rails as smooth as possible to reduce what we now call rolling friction, and he must have been mystified by a controversy that he could see was trivially easy to resolve.

Worse was to come. At a British Parliamentary hearing in 1825, his evidence in support of a Bill to allow the Liverpool to Manchester Railway was labelled 'trash and confusion', and he was personally reviled as 'an ignoramus, a fool, and a maniac'. The defeated Bill was resubmitted and passed some months later, thanks largely to staunch support from Mr William Huskisson, a Liverpool Member of Parliament. It was tragic that he was fatally injured at the public opening of the railway on 15 September 1830. Apparently, eight trains were involved in the inaugural run, and the opening ceremony was attended by many thousands. During a scheduled stop for water, Huskisson left his train to speak briefly to the Prime Minister, the Duke of Wellington, and after shaking his hand to mark the resolution of their recent differences, he rushed back to his train only to fall under another train—the famous *Rocket*—which severely crushed his legs. George Stephenson must have been as horrified as anyone, but in a dash for medical help he personally drove the train carrying the dying statesman for 15 miles at an average speed of 36 m.p.h., or three times the speed that many expert witnesses to the Parliamentary hearing had previously described as preposterous. Unfortunately Huskisson died of his wounds the same day.

The lack of a theoretical framework almost certainly contributed to the intensity of the opposition to Stephenson's work. Such concepts as heat, temperature, work, and friction were poorly understood at the beginning of the nineteenth century, and Stephenson's ideas were decades ahead of his time. It was not until the middle of the century that the work of James Joule and Lord Kelvin in particular, both British scientists, led a German, Rudolf Clausius, to extend Carnot's work and to formalize the first two Laws of Thermodynamics. Like maps of previously uncharted territories, good theories inspire confidence and give a sixth sense to perception. Those who came later could thus explain why the early steam engines were so inefficient, and point clearly to the ways in which they could be improved.

Meanwhile, Stephenson would not have been holding his breath. He was the most practical of men, and his main concerns were that solutions to the problems he tackled would be effective and economical. But these talents are not exceptional. Stephenson also had strategic vision, and the supreme determination to match his intuitive insight. His eventual success in attracting investors rapidly inspired others, and it launched one of the most breathtaking privately funded construction programmes ever seen, which rapidly transformed life in Britain, Europe, North America, and eventually almost everywhere. His financial success was also considerable. When he died in 1848, his personal bequests to his son Robert, who had followed in his father's footsteps with distinction, made him a sterling millionaire according to their biographer Samuel Smiles. Robert Stephenson was the first engineer to attain this nominally happy state, which, when the forces of inflation are taken into account, might just make him a dollar billionaire in today's money.

George Stephenson also developed extensive innovations in mining, and civil engineering generally. His genius was no less inspired than that of many great scientists, and few scientists can have made a greater contribution to human endeavour. Stephenson's contribution was, however, to the sciences of the artificial (see Chapter 4). He did not try to persuade Nature to give up more of her secrets as natural scientists do, but set out to devise new ways of using those he could see she had already released.

In the musical world, the distinction between composers and performers is well understood. The creative works of the great composers may be performed by many generations of great artists who can deservedly claim credit for the skill and quality of their individual interpretations. Indeed, the status and respect accorded to the performing artist are often as high and sometimes exceed those given to the composer.

The scientific world is not so well understood. The industrialists who work to develop and market competitive products rarely receive anything like the same accolades as are awarded to the originators of the discoveries underpinning their work, even though the levels of creativity and determination necessary to achieve and maintain a competitive edge can match or sometimes exceed those that lead to the initial discovery.

Several industrial scientists have won Nobel Prizes. For example, John Bardeen, Walter Brattain, and William Shockley won the Physics Prize in 1956 for their discovery at the Bell Telephone Laboratories in New Jersey of what we now call the transistor, which laid the foundation for the electronic revolution of the past few decades. At the same laboratory, Arno Penzias and Robert Wilson won the Physics Prize in 1978 for their discovery of the cosmic background radiation that is thought to come from the dying embers of the Big Bang that seems to have created the Universe some 15 000 million years ago.

More recently Georg Bednorz and Karl Müller won the Physics Prize in 1987 while working at the IBM Zürich Research Laboratory. They discovered that certain types of ceramic materials, not normally noted for their electrically conducting properties, became superconducting at much higher temperatures than some metals. The ability of a metal to conduct electricity usually increases as the temperature is reduced, but at very low temperatures the electrical resistance of some metals suddenly vanishes without the slightest trace. This remarkable phenomenon was discovered in 1911 by a Dutch scientist, Heike Kamerlingh Onnes, but it was not until almost fifty years later that John Bardeen, who was then at the University of Illinois, came up with an explanation and shared another Physics Prize with Leon Cooper and Robert Schrieffer in 1972. We all become sluggish if the temperature approaches zero on either the Celsius or Fahrenheit scales. The Kelvin scale has a much more fundamental reference point called absolute zero at $-273\,°C$, at which Nature switches everything off in absolute apathy. Superconductivity in metals is often found at temperatures some 10 or 20 degrees higher than this frigid fiduciary. These temperatures are very difficult to reach, and so the phenomenon had limited applicability. Bednorz and Müller's discovery, however, raised the prospect that materials might be found that are superconducting at 'room temperature' ($+15\,°C$). In 1991, superconductors that would operate at 'room temperature' in Vostok, Antarctica ($-117\,°C$) were found by Peter Edwards, who is a Venture Researcher, and his colleagues at Cambridge. In 1993, the prospects for more hospitable environments seemed quite good.

Many other major discoveries in the material sciences have been made in industrial research laboratories. Irving Langmuir is one of the most celebrated examples. He spent almost his entire scientific career, from 1909 to his retirement in 1950, at the General Electric Research Laboratory in Schenectady, New York, where he made a vast range of contributions in valence theory, catalysis, electric discharges, and atmospheric physics. He won the Nobel Prize for Chemistry in 1932.

It is nevertheless the case that basic research within the industrial sector has usually been carried out only by permission of an imaginative or far-sighted manager, director, or owner. Occasionally, it has been done in secret, 'behind the fume-cupboard', or disguised as something more prosaic to please the accountants. Sadly, it has now become very difficult for industrial scientists to pursue their own curiosity-led research, particularly in the past few years, as the quest for competitiveness has led companies to shorten their time-horizons, thereby almost eliminating another source of diversity. It should be noted, however, that many large Japanese companies are trying to resist these trends.

In many countries, companies have a statutory obligation to use their shareholders' investments to the best advantage. Industrial scientists must therefore be prepared to subordinate their personal freedom to the achievement of the company's objectives. In the earlier chapters we have discussed the role of freedom in making scientific discoveries, but freedom is not of course enough and among many other factors luck plays its part, in a form often referred to as 'serendipity'.

'Serendipity' is not derived from some ancient root as most words are, but was coined in 1754 by Horace Walpole, the literary son of the long-serving British Prime Minister Sir Robert Walpole. The word was inspired by a Persian book of fairy tales, *Three Princes of Serendip*, whose heroes were 'always making discoveries by accident and things they were not in quest of'. Serendip was the once-popular name for the island we now call Sri Lanka. Walpole suggested that the word 'serendipity' should be used for the faculty of making happy discoveries by accident, wherever they take place.

Scientific discoveries, however they may be achieved, do not directly lead to industrial products. The road to the marketplace can be very hazardous, especially for those potential opportunities that are science-led with no obvious markets. Success usually comes only when high levels of ingenuity at every stage of development, manufacturing, and marketing are matched by the commitment to succeed. In the past few decades the Japanese have been conspicuously successful in enterprise of this type, and following Horace Walpole it seems appropriate to coin the word 'japanity' to mean the ability to transform concepts into profitable business.

George Stephenson had this ability too, and so have many other western individuals. There is good evidence therefore that japanity will be no more eponymously confined than its delightful precursor. However, for manufacture in general, the compound annual growth in output per employed person during the period 1950–87 was 8.0 per cent in Japan compared with 2.8 per cent for the UK and 2.6 per cent for the United States; and to remove the distortions introduced by the Second World War's devastation, the equivalent figures for the period 1975–85 are 6.9 per cent p.a., 2.9 per cent p.a., and 2.2 per cent p.a. for Japan, the UK, and the United States respectively. Other nations, notably Korea in the east, and Italy and Germany in the west, have performed almost as well, but there seems ample justification for the proposed accolade.

Many have claimed that Japan's outstanding performance owes much to her long-established and strict social structures that tend to encourage teamwork and inhibit individuality, and the consensus seems to be that there are few lessons to be learned from them that would be relevant to the west. There is no question that many Japanese companies have wisely selected policies based on the strengths of Japanese cultural characteristics, but few nations have given as much priority to the systematic study of development, production, and marketing, and to their subtle interrelationships, which must be understood if these processes are to be coherently integrated to achieve a successful output. Inasmuch as the Japanese have turned these arts into sciences, it would be remarkable if many of their achievements were not universally applicable.

But the west has its strengths too, and for more than two hundred years its scientists have provided a scintillating series of discoveries and inspired industrial growth. The nineteenth-century French scientist

Louis Pasteur was among the most prolific. He was also one of the first to appreciate the value of a full and mutually stimulating relationship between research and industry. Among many other things, he drew inspiration from, and in turn eventually transformed, the brewing of beer, wine, and vinegar, and the production of silk with his extended work on micro-organisms and their metabolism.

Michael Porter pointed out in his book *The competitive advantage of nations* that the Japanese '. . . have a long tradition of adopting parts of other cultures. There is a high level of respect in Japanese companies for strong rivals and a lack of technical arrogance or concern over authorship.' In this spirit, therefore, it would seem that the east and west have complementary skills, and that if we could as a matter of routine combine the fruits of freedom and serendipity on the one hand with japanity on the other, there would be prizes in such profusion that even a latter-day Pasteur might be impressed.

It was not until about a hundred years ago that industry began to move away from a virtually total reliance on the flair, vision, and determination of individuals and to start building institutions that would systematically search the scientific crop for ideas that might enhance the performance of current products or lead to new ones. It might have been expected that overall progress in this direction would have been greatest in those parts of the world where the scientific harvest was richest. Indeed, towards the end of the century there was widespread recognition by German industry and government of the potential strength of their position, and there were isolated examples of far-sightedness in other countries. It was not until the First World War, however, that the need for industrially sponsored research and development came to be generally recognized in Britain and France, as it had been in the USA a decade or so earlier. There was steady growth through the 1920s and 1930s, but it took the dramatic technological successes of the Second World War in such areas as radar and atomic weapons to launch the industrial research laboratory towards its present importance.

Nowadays, almost every technological company employing more than a few hundred people will devote some of its resources to activities such as quality control, trouble-shooting, or the planning and preparation of new products. These activities may not themselves immediately contribute to a company's output but they are nevertheless essential to longer-term survival. A company that is expanding, no matter at how small a rate, is usually more profitable and much easier to manage than one that is stagnating or contracting. The problem is that the greater the number and range of products leaving the factory today, the more substantial will be the rod that will be used to beat the company's back tomorrow in order to encourage the achievement of still higher levels of performance.

Companies will, however, find little to guide them when they come to decide the proportion of their resources that should be allocated to the maintenance of a competitive edge. Many companies take the expedient of ensuring that their own research and development expenditures are not conspicuously different from their major competitors. Shareholders and other financiers, at least in the west, tend to regard company research and development as a cost to be minimized rather than as an investment in future prosperity, and their wrath can be swifter and more terrible than the customer's. Thus, within each industrial sector, research budgets tend to have the same broad relationship to company size.

Academic research is often performed for its own sake, but in industry the primary purposes of a research laboratory are mainly to look after a company's technological health and to provide a degree of insurance against unexpected developments elsewhere.

The laboratory's performance will be influenced by a number of factors, but perhaps the most important of these for a company of appreciable size is the calibre of the research director. Smaller companies often manage without one, or the post may be combined with production, marketing, or sales. In a large company, the research director must be able to win and maintain the confidence of the other directors, and eventually the board, whose members will be predominantly non-technical. He (even today, research directors are almost invariably male) must persuade the board that the company's research and development facilities are necessary and give at least as much value for money as any other part of the firm. These are very much on-going responsibilities, and few research directors today are given much respite from the virtually endless reviews of their function. But these persuasive talents would be wasted unless the research director is also able to inspire, motivate, and manage the often considerable numbers of scientists and engineers responsible to him, and who must deliver on time and on budget if he is to keep his credibility, and therefore his job.

These tasks might appear to be little different from those of other directors, and that is generally the case. But for the research director, the commodity to be delivered requires a deep understanding of some facet of Nature's behaviour, and of the ways in which this knowledge can be manipulated and controlled: Nature, unfortunately, is notoriously resistant to being managed.

In 1909, when Irving Langmuir was persuaded to join the General Electric Research Laboratory by its first director Willis Whitney, one of the company's most important problems was the short life of the recently developed tungsten filament lamp. Based on their considerable experience, Whitney and the rest of his staff believed that the best line of attack was to improve the vacuum in the lamp, and consequently the achievement of very low pressures was one of their greatest strengths, a facility which had impressed Langmuir when he joined them for a summer job in 1909: a glorious 'summer' that lasted until 1950.

Langmuir's biographer, Albert Rosenfeld, points out that the sign on Whitney's office door read, 'Come in—rain or shine.' Langmuir's first proposal to Whitney was that he should tackle the lamp problem by studying the effects on the tungsten filament of the very gases that his colleagues were doing their best to eliminate. Every maverick needs a patron, and Whitney showed that his mind was as open as his door by giving Langmuir his blessing, which he maintained even though for some time the work seemed to be going nowhere.

Langmuir's enquiries were wide-ranging, but he never lost sight of his primary goal. For example, he discovered that hot tungsten filaments could break the interlocking bonds that hold molecules together, and he was the first person to study hydrogen in its violently reactive atomic form. His work opened up a rich vein of possibilities that led eventually to new insights into catalysts; into plasmas—that high-temperature state of matter in which electrons and their positively charged nuclear partners are excited into such a frenzy that they have no time for stable, long-term relationships; into the reactions between surfaces and gases in general; and inevitably to the production of better lamps. Langmuir realized, as other pioneers have done, that difficult problems are rarely solved by staring them in the face and trying to force submission, step by reasonable step. Indeed, in this instance reason was driving the laboratory expensively in the wrong direction.

One of the faults the laboratory was working on was the progressive blackening of the bulb. The conventional wisdom was that the blackening was somehow caused by water vapour: hence the need for very low pressures. But Langmuir proved that water vapour could not possibly be the culprit: when the gas was bubbled into the lamp through liquid air, which is some 190 °C below the freezing point of water and its vapour, the blackening was unaffected. Soon after, he found that tungsten was being evaporated from the hot filament itself, which thanks to the high vacuum, could shoot across the empty space and condense on the bulb, eventually blackening it. He reasoned that if he could reduce the evaporation rate, not only would the filament last longer, but the lamp would burn more brightly. The answer was to surround the filament by a blanket of relatively inert gas—nitrogen was used at first—from which evaporating tungsten atoms would bounce back to the filament.

When Langmuir told Whitney about this beautiful result, which implied, of course, that the lab had been on the wrong tack for some time, Rosenfeld respectfully relates that Whitney's response was to tell Langmuir that he was dreaming. It is likely that Whitney had a great deal more to say than that, but he encouraged Langmuir to develop his ideas and Langmuir went on to file a patent in 1913. The best blanket gas turned out to be argon, and the lamp conceived by Langmuir became the ubiquitous light bulb, the design of which has survived virtually unchanged to the present day.

Ideally, therefore, a research director should know what is known within the appropriate industrial sector but should not be its prisoner. He should also appreciate the potential value of that knowledge to present and future products; be aware of which new knowledge might reasonably be given up to a determined but cost-effective research programme; and be able to recognize the types of knowledge that would probably not be acquired by directly commissioned research whatever resources might be thrown at them. Not surprisingly, it is rare to find all these disparate scientific and business-like talents within a single individual, and those who would claim to have them would probably not be averse to creating the impression that if required they could walk on water too.

Those intent on an industrial career must decide its timing as well as choosing a suitable company. These decisions may be related. A first-degree course includes almost no research training, and if you decide to go into industry, your choice of company will be influenced by the types of training and career development on offer. Some companies, especially the smaller ones, prefer fresh graduates with a good general education rather than those with specialist training because it gives them greater flexibility in their use of personnel. Indeed, almost all Japanese companies, including the largest, recruit their scientists and engineers mainly at the first-degree level, and take the view that their training *starts* when they join the company. Thus, Japanese companies tend to place little emphasis on the field of your university or college training. Your career might easily develop along quite different lines from those you have studied, and might also change direction more frequently than is the tendency in the west.

The preference of some of the smaller western companies for new graduates may also be influenced by their general lack of confidence in the ability of the academic sector to provide other than basic training. They might prefer that you were not exposed for too long to what they see as the bad habits of academia: a perceived lack of sensitivity to costs, deadlines, and an inclination to generate problems rather than solutions. Indeed, the novice will soon become aware of the mistrust and misunderstanding that pervades academic–industrial relationships. Many academics believe that industrialists are the ultimate in materialists who know the cost of everything and the value of nothing; whereas many industrialists believe that academic life is a bit like sex: when it is good it is marvellous, and when it is bad it is still very nice.

Large companies in the west tend to recruit people with research experience because in general they are looking for specialists who can readily contribute to the company's fields of interest. Indeed, from the graduate's point of view, postponing entry until after a master's degree or a doctorate can be a good idea if you set aside the obvious short-term financial and other material disbenefits. The academic environment can in some respects be more forgiving of beginners, and should give you an opportunity to develop a broader base to your research expertise in a somewhat more relaxed atmosphere than an acutely cost-conscious industry generally allows.

The academic sector is not immune from financial concerns of course—far from it, as we have discussed in the earlier chapters. Traditionally, academics needed to finance marginal costs only; that is the actual costs of doing the proposed research: the equipment and research staff, etc. All other costs, including salaries of tenured academic staff such as professors and lecturers and their accommodation were usually provided from general academic budgets. This simple situation is changing, however, and academics are now having to take responsibility for finding a progressively larger proportion of the total costs from peer-reviewed sources. In industry the budget the research director fights for has always had to make full provision for all staff, including administrators, gardeners, and those guarding the gate, as well as those costs directly relating to research and development, and every brick and piece of furniture or equipment in every building.

Most companies take care to protect new research staff from the full rigours of competition until they find their bearings. The induction programmes give time to take stock as they progressively introduce the recruit to the company's customs, ethos, and objectives. A newcomer is rarely thrown in at the deep end; career development—work, training, and promotion, etc.—is indeed given a great deal more attention than it would generally receive at a university or college. Many companies carefully monitor the progress of their young staff through the first decade or more of their careers, especially for those whose early performance seems to indicate that they are destined for senior positions.

Although there is ample scope for individuality and personal ambition, a typical industrial company tends to manage its staff as valuable assets, and in turn turn staff are expected to become organization men and women as soon as possible. The challenge to the individual therefore is to contrive to work in the mainstream areas best suited to one's own talents, and to avoid projects that seem to be going nowhere. These will be familiar problems to people in any walk of life, but inasmuch as scientists are driven by a craving to understand the unknown, those who turn to industry today must curb their appetites for research, and learn to satisfy the hunger of their souls with whatever requirement for new knowledge happens to be on the menu in their part of the company.

But industrial scientists have degrees of freedom often denied to their academic colleagues. Industrial products must function in the everyday world. Industrial scientists therefore do their best to ensure that their enquiries are not artificially restricted by the constraints often imposed by academic disciplines. Thus, a typical industrial research team might be made up, for example, of chemists of every academic hue working alongside physicists, mathematicians, and engineers, and their efforts harmoniously orchestrated by a sensitive and perceptive team leader. This does not mean that the team's enquiries will be free-ranging. On the contrary, their leader will impose tight control on the types of questions that can be pursued, and must be convinced that the answers will be worth having before any work is authorized. Today's Langmuirs may not be so lucky therefore. The more they are denied, of course, the more expensive will be a laboratory's failures.

These dimensions of industrial freedom can therefore often be illusory. Industrial research laboratories tend to be strictly hierarchical, and research and development programmes are ultimately subject to the approval of the research director and his senior staff. Some directors take care to encourage the flow of new ideas from junior people, but it is not easy. Just as junior or non-commissioned officers can influence generals, they need to be very sure of themselves before they stick their necks out, and have the skills necessary to pilot their proposals through the layers of management that usually separate bench scientists from research directors. Some directors pay only lip service to keeping these channels open since they believe that to accept ideas emerging from the ranks would be to concede weakness in their own management. When money is tight it is all too easy to close these channels completely.

Whatever the limits on personal freedom, those who choose an industrial career are usually happiest when they are making things happen, and helping to produce goods and services that will sell, because they are cheaper, or better, or more exciting, or otherwise more attractive than is offered elsewhere. There is much personal satisfaction to be gained from playing a part in the production of successful products. Companies also encourage their more creative staff to publish whenever commercial considerations allow. Publications not only help to maintain morale among laboratory staff, but also help to attract applications from the best graduates, as a company's reputation for excellence is extended among a wider audience than its direct customers.

The productivity of an industrial laboratory will indeed depend crucially on the levels of satisfaction its employees enjoy, which in turn will be strongly influenced by their perceptions of the qualities of leadership imposed upon them. Like the rank and file in any walks of life, bench scientists too have their mordant humour. BP's headquarters were generally known as 'big panic house'. In 1990, a memo purporting to come from managing directors at Rolls-Royce stated that: 'in the interest of economy the light at the end of the tunnel will be extinguished until further notice.' In contrast, a popular slogan among the scientists and engineers at the Sony Corporation in Tokyo in that year, was 'BMW: beat Matsushita whatever.'

Leadership can of course be efficient only if it is well informed. Sadly, many a company's senior executives have only passing acquaintance with technological issues and tend to be uncomfortable with them. In Britain, and in some other countries, science and technology have not traditionally been regarded as essential components of a well-grounded general education. Thus, the late Robert Belgrave, a senior executive at BP, and an arts man who was always sensitive to the two-cultures problem, addressed a management conference in 1980, and said that in future all graduate recruits should know *both* the difference between federal and confederal, and the laws of thermodynamics. He did not get his way—perhaps because there would have been a dearth of eligible candidates.

The aversion of many company boards to technological issues creates difficulties for all concerned. Research directors are rarely given the time they need to cultivate commitment and confidence among board members, either because boards rarely give intensive and extended consideration to technical issues—once or twice a year tends to be typical in many western companies—or perhaps the technical people are not board members. In their turn, boards can be placed in the embarrassing position of either having to accept the recommendations of their technical officers in full, or if they do not, to risk taking decisions that may be technologically irrational or ill-informed. When the competition for resources is intense, it is inevitable that some of a board's directives will mystify the *cognoscenti* and evoke ribald responses from the ranks.

Research directors may have to put up with the occasional unmerited mauling, and they too will have their critics, ribald and otherwise, but their privileged position usually gives them enormous power and influence over a company's technological affairs. There are a diversity of approaches to company research funding, but in general, each specific division of a large company (oil products, telecommunications, or pharmaceuticals, say) decides each year the level of research and development support it needs and can afford from its own resources. These decisions include the scientific programmes to be undertaken, the number of man-hours to be spent on them, and the objectives to be achieved. Decisions will usually have been taken in full consultation with research managers but once they are agreed there is little room for manoeuvre.

The research director, however, often has a largely discretionary budget that may be 20 or 30 per cent of the total and is intended to cater for the needs of the corporate body as a whole in such general areas as information handling, in which every part of the company would have an interest, and to provide early warning of new technological opportunities. In a large company, the corporate research budget may amount to several tens of millions of dollars, funds that government laboratory directors, or academics at any level could usually only dream of controlling. Academics, of course, do not often command large battalions, and although government laboratory budgets can be large, they tend to be subject to intense bureaucratic control, so that the director has little scope for personal initiatives.

The industrial research director, however, can usually draw on the corporate pot for any purpose that he can convincingly argue might be in the company's interest. This obligation is rarely a difficulty, as the power that normally goes with the job of being able to offer the glittering prizes of patronage and largesse can be inspirational, and research directors are not normally noted for their reticence. Needless to say, corporate budgets are subject to the most intense scrutiny of all. Their diffuse relationship to profitability makes them an obvious target for the 'bean-counters', and can test the patience of the most far-sighted patrons. Research directors therefore need all the stimulants they can get if these boardroom battles are not to become wars of attrition.

In more relaxed times than those of today, bench scientists were usually allowed to develop their own lines of research provided they were relevant and took no more than say 5 or 10 per cent of their time. That flexibility has almost entirely disappeared in the west, although the Japanese in particular have very sensibly preserved it. In the west in general, therefore, the creative bench scientist's best hope is to persuade the research director to release some of his corporate funds.

In 1979, BP's Dr David Graham was playing host to Ignacio Layrisse, a young student from Petróleos de Venezuela. Graham's expertise is in colloids, materials in which very fine particles or drops are dispersed in another fluid—a liquid or gas. Colloids have long been known, but it was not until the 1860s that Graham's namesake Thomas, a Scottish chemist, properly identified such materials as gelatin, milk, and albumen, coined the word 'colloid' from the Greek word for glue, and went on to put colloid science on a firm footing.

Layrisse had come to BP's main research laboratory at Sunbury-on-Thames in England for a few weeks to seek out ideas for dealing with Venezuela's vast heavy oil deposits. Oil is normally found dispersed at high pressure in certain types of rock, usually far below the Earth's surface. If its viscosity is low enough, the oil can be brought to the surface and taken off through pipelines or by ships for refining into petroleum products. The world has reserves of some 1000 billion barrels (or about 100 billion cubic metres) of recoverable oil that can be transported in these familiar ways.

Exploration for hydrocarbons such as oil or gas is a chancy business. After extensive geological surveys, sites might be found where there are all the signs that hydrocarbons are down there somewhere, but they may be as deep as three miles or so. The only way of confirming the putative discovery and evaluating its size and potential is to drill into it. Even if hydrocarbons are actually there, the uncertainties are such that the reservoir might be missed with the first few stabs. Most drillings, perhaps 90 per cent, turn out to be 'dry'; that is, ironically, they yield only brackish water. During the 1980s, BP discovered at Mukluk what might possibly be the world's most expensive water wells beneath the Alaskan Beaufort Sea. After an expenditure of some $600 million, the exasperated explorers finally concluded that they had drilled possibly a few million years too late, and the oily deposits that had once been there had seeped away.

Even when an oilfield has been found, the explorer's celebrations can only begin when it turns out that the oil can be extracted and taken away for refining into saleable products. Thus, fields containing total reserves of some 4500 billion barrels of oil have been found in various parts of the world, but much of the oil is thick and sticky. Extraction (the term generally used is 'recovery') is not a problem if the rock is porous and the oil is hot and fluid enough to allow it to ooze to the surface under the natural pressure in the reservoir. For relatively light oils, these natural processes (which are usually called 'primary recovery') might allow a reservoir to yield up to about a third of its bounty, but for the heavy oils, the proportion drops to less than a tenth. Thereafter, progressively more expensive secondary recovery processes, such as injecting the reservoir with water or steam, have to be used. Even for light oils, recovery levels rarely exceed a half of a reservoir's potential.

About a third of the tarry deposits lie in Venezuela, in the Orinoco Belt. The oil is found at depths of only a few hundred metres, but since it is hot (70° or 80 °C) the initial pressure is sufficient to bring the oil to the surface when the reservoir has been tapped. The enormous problem was what to do with it when it cooled and virtually solidified.

In everyday life, soups can be thickened by adding flour, but anyone who did this directly would end up with a lumpy unappetizing mess. Flour and water can be emulsified, however, and, as every cook knows, a little hot oil or fat mixed with the flour before adding it to the soup will do the trick. In the language of the colloid scientist, the cook has found a surface-active agent, or surfactant, which encourages the flour to enter into an intimate relationship with the water rather than sullenly sitting on its surface as it would normally do.

Layrisse realized that collaboration with Graham could be very valuable, and stayed for eight months. As they worked together, they quickly found, as others elsewhere had done, that the heavy oil and water could be emulsified into a colloidal solution with the help of some detergent, but the emulsion was a poor sort of thing and tended to be as lumpy as an amateur cook's soup. This familiar and frustrating outcome had been found elsewhere, but Graham thought that his expertise on oil and water's abstemious relationship, and on surfactants in general could bring these reluctant partners together, and might be the key. But first the legal relationships between BP and Petróleos de Venezuela had to be sorted out, and as ever, lawyers will not be hurried.

The agreement was signed in 1982, but Graham had to find a BP sponsor for the work and neither the Oil nor the Exploration divisions were sufficiently interested to pay for it. Undeterred, Graham went to the research director, Professor John Cadogan, who luckily was quickly convinced of its corporate importance and potential. Layrisse had now returned, and Cadogan allowed Graham to devote 5000 man-hours a year to the problem for the next four years.

For almost two years the work did not go well. The conventional route to emulsifications is to apply the highest possible levels of shearing force to the would-be mixture, together with a little of the most imaginative cocktail of surfactants. But it simply did not work. Whatever he tried, and however large the shear, they always ended up with the same gooey glop. However, in a flash of inspiration of which Irving Langmuir might have been proud, he decided to fly in the face of convention, and *reduce* the shear. Low-shear mixers are not normally found in a high-tech lab, so he used the cheapest food mixer he could buy.

He had realized that just as say butter and flour can be combined into a creamy emulsion only by first using a slow folding action, the same idea might apply to the heavy oil and water. It did. The resultant emulsion turned out to be as runny as ink, and just as stable.

Graham demonstrated his work in 1990 at a Royal Society Soirée in London. These are grand, after-dinner affairs, at which fifteen or so recent discoveries are demonstrated to an invited audience including many of the great and the good, with everyone dressed in their formal finery. Each demonstrator has a stall, and the wine-sipping guests glide among them, pausing whenever something takes their fancy.

Graham gave a demonstration to myself and my wife; others quickly joined us. His stall was conspicuously different from the rest in that it looked more like a family kitchen than a scientist's laboratory. There were food mixers and plastic bottles of washing-up liquid next to a kitchen sink, vying for our attention with the latest in computers, lasers, and other high-tech wizardry offered by the other exhibits.

His demonstration started with a bowl of Orinoco oil, black and stolid as he turned it upside down, just to prove the point. He added water from the tap, followed by a squirt of detergent. The mixer then started to work, very slowly at first, but within minutes the languid liquid came to life, and threatened to ruin the Royal Society's carpet.

Cadogan had by now joined our party, and heard Graham extolling the technical virtues of the process, which he said was now a commercial initiative. Orimulsion, as the fluidized heavy oil is now called, is indefinitely stable, may be pumped down pipelines, and because of the high proportion of water—some 30 per cent—and the fine droplet size, the fuel burns very efficiently. Like other fossil fuels, heavy oils contain sulphur, but economic ways of extracting it from smokestacks already exist. The work was a *tour de force*. It elegantly demonstrated that, in spite of a widely held opinion today, great discoveries do not necessarily need expensive equipment, and can still come from the expert use of the stuff between the ears. I could not resist congratulating him for a magnificent piece of work.

Cadogan, however, was not to be upstaged. While David Graham was still demurely basking in the little group's acclaim, Cadogan walked across, took Graham's blushing cheek between finger and thumb, waggled it and said, 'Yes, but who was it who persuaded the managing directors to go ahead?'

On the face of it, this was a most surprising thing to do to a senior colleague. However, one of the most frustrating aspects of an industrial bench scientist's career is that many more discoveries are made than can possibly be taken forward. If the group leader or, ultimately, the research director does not support a particular piece of work it may not see the light of day. In these circumstances, patents might not even be applied for if the expense could not be justified, and there might be commercial reasons preventing publication of the results in the scientific literature. Even if the director supports development, he too might be turned down by a committee or by the board. He will not wish to see that happen too often.

Cadogan had indeed promoted Graham's work with some vigour. He persuaded the relevant managing director (Robert Malpas) to give his approval, and eventually a new joint venture was set up with Petróleos de Venezuela (BP Bitor) specifically to exploit the discovery. Graham, however, had unfortunately omitted to emphasize his research director's vital role in all this, and Cadogan had merely made sure that he would not do so again.

Graham's discovery could turn out to be crucial to the world's energy future, but it is not of the type for which Nobel Prizes are awarded, since no new science is involved. Nevertheless, Graham showed the flair, determination, and courage that is characteristic of great scientists, and it is a pity that there does not exist an award comparable in status to the Nobel Prize that recognizes major industrial contributions, not only for the benefit of the individuals concerned, but for the stimulating and inspirational impact it would have on industrial science and technology.

This story could have turned out quite differently. One of the most ubiquitous obstacles found at every turn on the rocky road to the market goes under the epithet—'not invented here'. As ideas progress from an initial concept through various levels of assessment and development to the more complex stages of manufacture and marketing, there are corresponding changes in the expertise required, and thus a progressive expansion in the numbers of people involved. Even in an ideal world there would be many pitfalls for the unwary, for even some types of abstract problems do not have neat solutions. But the everyday world is far from ideal. Not only are the materials that go into products or processes less than perfect, but people are involved. The complexities that people generate frequently lead to delays in getting new ideas accepted, and this is often because they have been invented somewhere else.

Not surprisingly, these problems are more acute the bigger the jump an idea or an embryo technology has to make; and the jump across the cultural chasm separating academia from industry is the biggest of all.

Some people, in Britain at least, hold the view that to transfer technology is a confession of a failure to develop the technology in the right place in the beginning. According to this view, it is difficult enough to transfer ideas from an industrial research laboratory to a plant owned by the same company. That barrier must be higher for ideas brought in from outside—those from a university, say—and their penetration will consequently be low.

But it would be a rare technology indeed that would grow like a plant in an isolated pot, however carefully it was nurtured. Rather, technological development has much more in common with the growth of saplings in a natural forest. They draw their nourishment from an environment that includes the forest floor, with all its fungal, bacterial, and insect residents, not to mention the birds and the bees, as well as a good supply of air, water, sunlight, and other essential nutrients. If a development is to be successful, therefore, it is essential that managers should create environments that encourage the inflow of ideas when the inevitable problems arise. All too often, however, necessity smothers invention, particularly if it comes from an unscheduled source.

Thus, those from outside a group might be reluctant to share their new ideas with others because they are afraid that they will get little credit if the ideas are taken up. On the other hand, those within the group might believe that if an imported idea fails for some reason they will get the blame. The acid of mutual suspicion can be very corrosive, and all too often new ideas can appear as threats rather than opportunities.

The very fact that the problems of transfer are often referred to in the context of 'not invented here' indicates that they are not very well understood. The authorship of most discoveries can be traced to an individual or group, and so if an invention has been made elsewhere, by George Stephenson, or by Irving Langmuir, say, those who would use the 'not invented here' label would therefore seem to believe that the problem is insoluble.

The difficulties have nothing to do with invention, of course. 'Incentive' should be the key word in that dreadful epithet, and with sensitive leadership the problem should never arise. It is ironic, however, that in a recession competition for resources between groups within a company becomes severe as the overall research budget gets squeezed, and leads to introversion as groups struggle to survive. It is difficult for managers to maintain an open mind and the free flow of ideas when the bailiffs are bashing at the door. Thus, as the need to be competitive becomes more critical, companies become more inward-looking and conservative.

None of this is inevitable, however, and industrial prospects will become brighter when leaders resuscitate their aptitude for positive thinking. A return to 'core' business, for example, as many companies in the west were doing in the early 1990s, is the same as concentrating on the things that have been done well in the past, and as long as such action is widespread, recession will continue. Conversely, when leadership begins to recognize the wealth of opportunities that will have accumulated even the worst recession will eventually come to an end.

7

Science and economic growth

So far I have outlined the ways in which the pursuit of science has evolved, and have tried to portray something about what it is like to be a scientist in at least two walks of life, namely the academic and the industrial, while hinting that the lot of scientists employed directly by government, say, or by other organizations is not radically different. The perspective has generally been that of the individual, and I have depicted the struggles a young scientist might have to bring ideas to fruition or to develop a career. In the remainder of the book, however, I shall try to look at the scientific enterprise from a much broader standpoint, that is from the point of view of people that are not necessarily directly involved in science or technology, but will ultimately come to enjoy, or have to endure the consequences of, these enterprises.

Throughout history, countless numbers of individuals driven by curiosity, or some other restless urge, have committed their own energy and resources to the improvement of some aspect of mankind's lot, and perhaps also to the cause of their own fame and fortune. For the past hundred years, however, the scientific enterprise has progressively become more institutionalized and very expensive. Nowadays, few scientists or engineers can finance their own research or development, and the responsibility must be shouldered by others if they are to bring their ideas to fruition. But when one set of people does something for another, there is always the question of why they should do it.

For the scientific enterprise, motives can vary enormously from the most enlightened altruism on the one hand to the desire to make money on the other, but whatever the motives, they will give little guidance to the benefactor or financier on the level of funds that should be provided. For charities, trusts, or similar organizations, the problem hardly arises: the funds available and the intended purposes are tightly controlled by such legal instruments as deeds of covenant and the like. Thus, the Wellcome Trust, for example, supports extensive research in medicine, a field that in general is well endowed by private sources.

Indeed, in 1991 charitable sources of funds for medical research in Britain exceeded those provided by the Medical Research Council, which is a government-funded body, and in 1993 the Wellcome Trust was the largest source of such funds in Britain. Other fields are not so well served in the UK, but the USA has a multitude of private sources, some of which can be very specific. The Robert A. Welch Foundation, for example, distributes substantial funds for research in chemistry within the state of Texas.

Industry and governments, however, together provide by far the major part of support for science and technology. Their motivations vary, of course. Industry needs to make profits for its shareholders, and science and technology can play an essential role in that process through the provision of viable and competitive products that can readily be manufactured and sold. Governments have much wider responsibilities in these areas, and usually will work towards achieving a well-educated population, a broad science base, and an infrastructure that encourages industrial production, international competitiveness, and economic growth.

Although objectives differ, the characteristic features of science and technology are the same everywhere. Nevertheless, in spite of this universality, there seems to be little agreement on how scientific enterprise should be nurtured, or on the levels of resources that should be allocated to such enterprise. Bearing in mind that science and technology have been the most powerful agencies for change in the industrialized world, and indeed can be said to have created it, the lack of agreement seems astonishing. Thus, it could be said, that having discovered a rare species of scientific goose, which in some magical circumstances can be relied upon to produce technological eggs of amazing variety and rare quality, industrial and governmental farmers seem not to have made it their business to find out how these geese might be best cared for, or how they might be stimulated into maintaining or even increasing the flow of their precious product.

The levels of financial interest also vary widely. In industry, as we have already remarked, many companies see research and development as a cost to be minimized rather than an investment, while others see it as essential to survival and prosperity. The commitment varies from a token or even none at all, which seems common among banks, financial houses, and service companies in general, to 10 per cent or more of the total sales revenue, particularly among electronics, pharmaceutical, or biotechnological companies. Thus, in such companies as General Electric in the United States, Glaxo in the UK, or Sony in

Japan, research budgets either exceed or are in the region of a billion dollars a year.

The research and development budgets of national governments are also highly variable. At a workshop held in London in the summer of 1988 to discuss the roles of industry and government in the funding of research, it was mentioned that there are some people in positions of influence who believe that the outstanding performance of British scientists in winning Nobel Prizes and the like over the past few decades has been a *result* of economic growth and prosperity, and not a cause of it. The implication was that such people regard science as a luxury that should be cut back when national belts have to be tightened. But the view that investment in the sciences is not linked to economic growth would generate as much confidence in many parts of the world as would, say, an airline pilot's decision to switch off an engine to save fuel because a storm was approaching. In the United States, for example, Ronald Reagan said when he was President in 1985 that 'No nation depends as much as we do on the science base.' And in Japan, the Deputy Director of the Science and Technology Agency said in 1988 that 'nations that fail to invest in research and development lose their vigour and prospects for growth and employment.'

The commitment of these and many other nations does not, of course, constitute proof that investment in the sciences leads to economic growth. Consensus is not necessarily reliable, particularly where there is uncertainty. Economics is one of the youngest of sciences, and the world is highly complex. Even terms in everyday use such as 'gross national product' can be measured in different ways and should be treated with caution when comparisons are made. National approaches to resource allocation vary considerably, even for tangible things such as highways or hospitals. Economists can credibly draw a wide range of conclusions from the same set of data, and so politicians have little to help or hinder them when priorities are being set.

For the scientific enterprise in general, which in our definition includes education, training, basic and other types of research as well as technological activity, the outputs are diverse and often intangible; and, despite a wealth of literature, there are no agreed ways of quantifying them. It is hardly surprising, therefore, that some governments prefer a 'not proven' verdict, and allocate a lower priority to investment in the sciences than others. But some nations have been confident over long periods of the value of investing in the sciences and have also enjoyed high rates of economic growth. Is it coincidence, or is there a real effect?

Interest in the relationship between scientific enterprise and economic growth stems largely from Robert Solow's work at the Massachusetts Institute of Technology in the 1950s when he introduced what he called a 'new wrinkle' in segregating variations in output per head due to technical changes from those due to the availability of capital. He was awarded the Nobel Prize for Economics in 1987 for this and other seminal work. The extensive literature on the subject concentrates mainly on the economic contribution made by technological change rather than that which might come from the scientific enterprise. The reasons for the relatively narrow focus are probably pragmatic; technological outputs are usually tangible. But although the economic potential of, say, a megabit electronic chip is more amenable to assessment than advances in such topics as quantum field theory, innovation can hardly be exploited efficiently unless, among other things, there is adequate understanding in the underpinning sciences, much of which has to be drawn from the scientific community rather than from an innovating organization. The importance and value to a national economy of investing in the sciences can be more easily and directly appreciated, therefore, if the exploitation of discoveries takes place in their country of origin.

Solow and others, in work focused mainly on the United States, seem to have demonstrated qualitatively that economic growth does follow technical change. The United States is a large economy, but there seems no reason in principle why that result should not be generally applicable. The 'wrinkle' offered here is to focus on the ways in which the intrinsic structure and characteristics of the scientific enterprise affects its gearing to the economy.

In a competitive world, the prizes tend to go to those with the highest levels of confidence and commitment. For science, however, as for other areas of human endeavour, these qualities are not likely to be sustained unless there is clear understanding of the enterprise, the objectives, and the risks involved. Unfortunately, the scientific enterprise seems to be widely misunderstood. Its success in the apparently effortless provision of a vast catalogue of exploitable opportunity over the years has led to the idea that science can be managed in much the same way as such other beneficial activities as the production of steel or oil, where supply and demand are governed by market forces.

But as we have discussed in earlier chapters, the scientific enterprise has its origins in our understanding of Nature herself. In common with other highly complex subjects, understanding in science should be *expected* to be incomplete, and to be merely an expression of the best that can be done for the time being. Newton's theory of mechanics, for example, is sufficient for the purposes he had in mind, but useless for understanding behaviour at the atomic level, for which the work of Einstein and others is required. It is inevitable, however, that the work of these more modern giants too will in time be superseded, not because they were wrong, but because there will be circumstances they did not foresee.

Nowhere are these characteristics more cruelly revealed than in industrial development, when new products or processes obstinately and expensively refuse to behave as they are supposed to. Innovation today stands at the apex of a broad pyramid of scientific endeavour, and if the foundations prove inadequate for the planned extensions, there is no alternative but to return to basics. The search for solutions may take the developers into the most unlikely areas of the scientific terrain: for example, a nuclear reactor development in Britain was held up until problems could be solved that arose from the presence of microscopic marine organisms in the cooling water. In addition, mere mortals are involved in all of this, and successful developments often owe as much to administrative and social skills as they do to science. Team-work plays an essential role, but a bad or ill-chosen word from a manager can destroy a team's spirit and turn enthusiasm into bloody-mindedness. It might therefore be as difficult to relate a successful technological product to a specific discovery as it would be to trace the origins of a forest fire to the first flickering flame that started it.

Nevertheless, the belief seems to be widespread that exploitation comes about in a simple sequence of events: research spawns discoveries, which in turn are assessed and developed, and eventually find their way to the market-place. This is, of course, a caricature of what actually happens. Occasionally it is not a bad likeness, as it was in the early days of the development of the transistor. But technology is usually the product of a complex web of relationships. For science-led initiatives, the spark of an original discovery, in electronics or in genetic structure for example, can affect a wide range of technologies. On the other hand, initiatives that are market-led may have to draw upon a wide base of scientific support covering many disciplines. Advances in the basic sciences in one field may stimulate understanding in another. And technological breakthroughs can cascade through many types of industrial and commercial activity. The changes produced can be dramatic. The banking, insurance, and other commercial sectors are totally dependent today on high technology in ways that could not possibly have been predicted even twenty years ago.

It should be expected, therefore, that the benefit to a national economy from investment in the sciences will come, not so much from isolated developments by innovating organizations, as from harnessing the technological pressure arising from the expansion of a technology and its progeny across a wide range of industrial sectors. The progeny in this analogy come from the gradual adoption and dissemination of a central idea (say, integrated circuitry), from imitation, and from improvements as companies try to make a product more attractive to customers through differentiation, or try to circumvent a competitor's patent. The speed of these diffusive processes varies among industrial sectors; it is strongly influenced by prevailing levels of communication and competition among industry, as well as by government investment in the scientific enterprise, and by specific initiatives like those in new materials, or biotechnology.

Those who would extract the maximum economic benefit from the scientific enterprise thus need to be aware of the pitfalls. At government level, perhaps the most important question concerns the extent to which a market can be relied upon to bring about levels of investment that are adequate to achieve national objectives. Government attitudes to intervention vary, but a government will watch how markets such as power generation are operating, and will take action only if there is clear evidence of market failure, which might in this instance take the form of a generating utility's failure to provide a national or regional requirement for which a government might assume an overall regulatory responsibility. Thus, a government might wish to be assured that the utilities are aware of the seasonal and daily fluctuations in demand, and will not be caught out if millions of people simultaneously switch on their kettles to make a drink during a break in a popular television programme.

In an advanced country, the power generating and distributing authorities will ensure that the causes of such spikes in demand are well understood, and can be met so that power will be there whenever it is needed. In general, therefore, markets can be expected to work more efficiently the better they are informed.

Markets in these contexts are normally taken as the means by which the exchange of goods and services takes place as a result of buyers and sellers being in contact with each other, either directly or through mediating agents or institutions. The concept of the market is not usually extended to science and technology, but their overwhelming importance to advanced economies seems to provide ample justification for doing so, for market failure in this context can affect national prosperity.

For science, however, the signs are that the market is not always well informed. For example, in the conventional market-place, supply and demand are treated even-handedly. In science though, to paraphrase Thomas à Kempis, the market proposes, Nature disposes. In the scientific enterprise, Nature always has the last word. There are very few areas of human endeavour that can be claimed to be well understood, and some technologies, such as a thinking machine or a cure for cancer, might not be available on demand in the foreseeable future whatever the levels of investment.

This is not to imply that governments can never rely on the mechanisms of the market in order to determine the right level of investment for the scientific enterprise. Indeed, since much technology stands on relatively firm scientific ground, the market, so to speak, would be expected to operate no less efficiently in these areas than it does in other fields of endeavour. But for the scientific enterprise, which, as mentioned earlier, we take to include education, training, research, and technology, the limitations of the market need to be understood. Those nations with the highest levels of scientific and technological literacy would therefore be expected to understand these problems, to see technological risk and opportunity as two sides of the same coin, and to obtain the highest yields from investment in the enterprise. This would indeed seem to be the case.

The commitments of national governments to civil research and development are the subject of intense debate among those involved. Some nations, notably Norway, Sweden, and Germany spent in 1990 some 1 per cent of gross national product on civil research and development, while surprisingly at the other end of the spectrum the USA, Japan, and the UK, manage only about half that proportion. These raw data should be treated with caution, of course. In Japan, some three-quarters of all research is funded by industry. The USA's economy is enormous, and in absolute terms the USA allocates far more resources to research and development than any other nation. In addition, the US and the UK governments, for example, commit substantial proportions of gross national product to defence. In the USA, much defence research and development, and production too, is contracted to the private sector, and to a lesser extent this practice is also found in the UK. Defence work tends to be classified, but the exacting standards demanded for defence equipment can contribute to maintaining a company's association with excellence and an infrastructure that encourages high performance. There is a downside to all this in that a tendency to produce 'gold-plated' products for defence might affect competitiveness; and the need for secrecy, coupled with the pressures stemming from the enormous scale of some projects, can and does allow corners to be cut, sometimes with dramatic consequences. The technological spill-over from defence seems, however, to be generally positive in some sectors, particularly in the aerospace and power industries.

Many critics nevertheless prefer to be economical with their caution, and the sport of statistic swapping is widely practised. As ever, politicians are as adept at choosing the figures that support their actions as scientists and others are at digging out the data that damn them. It is the case, however, that national allocations for research and development are nowhere near adequate to satisfy the demand, as we foreshadowed in Chapter 1. The shortfall is impossible to quantify exactly, but it might be by a factor of five or more in some areas, judging from the current rates of rejection of academic research proposals. In principle, this is not a new problem, for scientists have always been able to spend more than their patrons can provide, but new factors are now creating additional difficulties that are threatening the future of the scientific enterprise. This will inevitably have economic impact.

In the UK, for example, it has long been recognized that since the nation contributes only about 5 per cent to the world research and development effort, it is impossible to pursue all fields of research on a national basis. It has therefore been decided that the UK must be more selective in the research fields that can be supported, and should seek international collaborations in fields that the country cannot support alone. Thus, as we have outlined, committees of the best scientists and industrialists have set priorities between fields, and funding levels have been set accordingly.

On the face of it, these actions would seem plausible for a relatively small nation to take. In the United States, however, Dr Frank Press in his presidential address in 1988 to the 125th annual meeting of the National Academy of Sciences entitled 'The Dilemma of the Golden Age' said:

The United States supports more research than Western Europe and Japan combined, and our system of universities, and national, and industrial laboratories is the envy of the world. Why then is our community in an unprecedented state of stress and internal dissension?

He went on to say:

The issues are funding levels and priorities. Our political leadership has no way of gauging the amount of resources necessary to maintain the strength of American science and technology. What it does see is that the inevitable competition for funds leads to conflicting advice from within the scientific community. It learns of caustic debates among scientists in our journals and in the press. And it sees issues at times framed simplistically, as in the arguments of 'big science', as embodied in the super-conducting supercollider and the genome sequencing project, and 'small science', as represented by scientists working alone or in small teams. Arguments over funding priorities spill over into intellectual attacks on the worthiness of one field of research by practitioners of another. We see confrontation and competition bordering on the unseemly between basic and applied work, between traditional and new fields, between modes of doing research, from the single investigator to centers. At a time when we should revel in dazzling progress in almost every field of science, this sniping and carping among scientists is disturbing and destructive.

Press concluded with suggestions for the criteria by which scarce resources should be allocated, and called on scientists to agree on what those criteria should be before Congress left them no choice.

The situation in the USA that Press so eloquently describes is no less serious elsewhere, and if the richest nation on earth concludes that it is unable to give full play to the creativity and imagination of its scientists, what hope is there for the rest? As the arguments by which these re-markable conclusions have been reached and the proposals for sub-sequent action are virtually identical for rich and not-so-rich nations alike, it would seem that the problems have little to do with resource allocation. Perhaps some other factors not yet considered by the author-ities are having a substantial effect?

These problems stem entirely from success, as has already been mentioned, and as the title of Press's seminal address indicates. That success has, however, been accompanied by changes in infrastructure, institutions, and policies for the governance of the scientific enterprise that I would argue are in some important respects becoming worn-out and inadequate to meet the demands now placed upon them.

The three great pillars on which scientific enterprise stands today are the scientific disciplines, the peer-review system for selecting new research (which, as we have already mentioned in Chapter 5, is really peer *preview*), and the priorities for dealing with the funding deficiency. Some of the effects that the changes in these structures are having on performance in the sciences have been outlined in earlier chapters. We now turn to their economic implications.

Scientists have always been driven by a need to be the first to do something or to make a discovery. In the past the struggle has usually been between an individual and the problem in hand. In circumstances where the posing of a new problem is in itself a discovery, a scientist could be reasonably confident that no-one else would be working on the same problem, a condition that most scientists would have regarded as essential. There have been many scientific controversies about who was actually the first to conquer a particular summit (such as that between Isaac Newton and Gottfried Leibnitz over the differential calculus) but the academic tradition has been to try to avoid direct competition and races to solve specific problems. Industrialists face different problems, of course, but their targets and deadlines are usually precisely defined, and it might not be the end of the world if they were not met. It can often pay to come to the market-place later with a product that is better, cheaper, and has wider appeal, or is marketed with more determination or flair than the opposition's.

For academic research, however, there are no prizes for second place. When I was a nuclear physicist working at the University of Alberta, I had collaborated with my colleague Tom Alexander for six months or so on a problem, only to discover when we had written our first draft of our paper that someone in the United States had just published an almost identical result. It was as if our work had been stolen, but by whom? Not by our competitor, for he knew as little about our work as we did about his. But the sense of loss remained. As some of our friends pointed out, we probably could have found a journal in which to publish our results, and then at least we would have another publication to add to our names. But the joy of discovery had gone, and we put our draft aside like yesterday's newspaper.

In our case, the duplicated work would have contributed only a tiny increment to the vast store of knowledge, but had we known that we were involved in a race we would have found something else to do. At the other end of the spectrum of importance, so too would Leibnitz, but he had no idea in 1775 that Newton had already taken the first inspired steps towards 'the Method of Fluxions' (now called the calculus) some ten years earlier. Newton, however, had a pathological phobia of publication, and he perversely delayed publishing the full proof of his claim to priority for almost forty years. By that time, however, the recriminations and accusations of plagiarism were already in full swing. The controversy continued, indeed, for many years after both scientists had died.

The growing tendency nowadays to limit funding to the priority fields, and the use of peer preview and consensus to select the best research within those fields mean that scientists in the advanced countries are increasingly being forced to work on broadly similar problems, and hence to compete. Communication among scientists is now very efficient and for the truly great problems there would almost certainly be no question today of a Leibnitz not knowing what his Newton had been up to. But the fact that scientific goal-posts, so to speak, have been created means that it is inevitable on the global scale that scientists will be pushed, in effect, into working on problems with objectives that might only be marginally different from those being pursued in other countries, and might sometimes be identical.

In any competition, most people will search for something that might give them an edge. For scientists, access to, or better still ownership of, the best equipment can transform their chances. The scientific equipment market is now substantial, and scientists are bombarded with advertisements for the latest techniques and equipment. The manufacturers parade their claims in the journals in glossy and expensive colour advertisements, using such snappy slogans as 'get your results faster . . . publish sooner . . . the power to do serious science . . . give wings to your research'—all calculated to give a hungry soul indigestion. Nevertheless, it is virtually impossible for mainstream scientists to ignore these claims, for their productivity would inevitably suffer if they did not have access to the latest hardware.

By increasingly concentrating their scarce resources in priority fields, funding agencies are severely limiting the scientist's range of interests. As these fields become more crowded, demands for more sophisticated and expensive equipment also increase, thereby compounding the funding deficiency problem. Furthermore, the belief is now widespread that all the 'easy' problems in every field have been solved, and that it is no longer possible to do so-called world-class research without using the latest and best equipment. Thus, scientific enterprise is not only losing its diversity: it is slowly but steadily being strangled by its own good intentions.

So what? The scientific community's worries rarely come to the attention of the general public. Many outsiders would equate calls for increased diversity with demands for more money, which of course are usually open-ended, and are usually ignored as special pleading. In any case, many industrialists, who after all are among the main users of the results of academic research, seem to believe that there is no shortage of new ideas; on the contrary, they would say that most large companies can afford to develop only a small fraction of their own scientists' creative output, so why should they be concerned about a shortfall elsewhere?

It seems astonishing and contradictory that on the one hand the funding agencies' ability to fund only a small proportion of the ideas coming forward is causing serious difficulty among the academic scientific community, while on the other, industrialists faced with an almost identical situation in their own organizations seem to interpret it as an embarrassment of riches. However, it can be seen that this predicament is an inevitable consequence of the changes that have taken place over the past few decades in the academic community's management of its affairs.

For industry, it is essential that business goals should be precisely specified, and successful organizations have always made sure that this is so. The academic funding agencies, however, have only latterly come to apply this policy to research. Some would say they have been forced to do so because of the growing funding deficiency. Nevertheless, by increasingly making use of the well-tried weapons of an industrial survival kit, such as a strict emphasis on priorities, and the stimulation of productivity, the academic funding agencies are acting in ways that are not always consistent with the nature of the enterprise over which they exert so much influence.

Traditionally, academic researchers have sought to extend the realms of understanding rather than solve problems. In some respects, scientists are like beachcombers in that they cannot know whether or not a particular pebble conceals something interesting until they turn it over. Beaches usually have a virtually unlimited supply of pebbles, and so beachcombers rarely run out of attractive options to explore. Similarly, if scientists are constrained to work within well-defined problem areas such as advanced materials, warm superconductivity, genome sequencing, drug-delivery agents, neurotransmitters, cancer, AIDS, and other disease-oriented research, their creative talents can usually identify a plethora of scientific pebbles that might be hiding something of potential value that might contribute towards the achievement of their selected goal, and it is almost impossible to assert confidently in advance that any one attempt might not be worthwhile. Faced with this profusion of possibilities, funding agencies have been forced to split the hairs of reason in their quest for criteria that will enable them to separate the few outstandingly good ideas they can afford to support from the merely very good they must find reason to reject.

Creativity is not restricted to academics. Industrial scientists also shower company executives with a variety of ways in which the company's goals could be met, and it is quite proper that an executive's future career should be influenced by his or her ability to pick the winners. But when academics resort to playing by similar rules and try to win additional funding from a government, say, on the grounds that industry will thereby have access to more ideas, they should expect the rebuff they usually receive, especially as the priority fields of academic research not surprisingly bear no small resemblance to the longer-term goals that industrialists themselves would like to achieve. Industrialists, however, should also realize that they should not be indifferent to the dire straits in which academic research now finds itself, since otherwise they will have to learn to live with the gradual deterioration of a feedstock that has served them well in the past.

At a meeting of the Anglo-Japanese High Technology–Industry Forum held in 1990, an afternoon was devoted to the theme of 'Time to Market' in three broad industrial sectors: advanced materials, automotive and aerospace; electronics and information technology; and biotechnology. A surprising outcome was that lead times for developing a new product to the stage at which it could be sold were now roughly equally long in each of these sectors, except perhaps for the software industry. The pharmaceutical industry has, of course, long been plagued by lead times of a decade or more, mainly because of regulatory requirements for testing and approval, but representatives from other sectors had a similar story to tell.

Even in electronics, which is often thought to be a rapid-response industry, it took the Sony Corporation thirteen years of sustained struggle throughout the 1970s and early 1980s, and \$200 million to develop an electronic imaging device and what was to become the now-ubiquitous video camera. The chances of success for a new product usually hinge on the commitment and vision of its champion: someone who *knows* that the project will work, who can enthuse the team, dispel doubts, and above all, can carry that conviction into the boardroom.

Enthusiasm alone will rarely be enough to convince the impassive custodians of a company's coffers that yet another tranche of umpteen million is required to overcome the latest snag, and that a delay of another year or two will not damage the project's prospects for profitability in the long run. In the case of the video camera, the problem was how to capture an image electronically in such a way that it could be faithfully reproduced with at least the same resolution and reliability that is typical of photographic film. Kazuo Iwama, the director of Sony's Research Centre, decided that this difficult objective could best be achieved using a charge-coupled device (CCD), an invention made at the Bell Laboratory in 1969 by W.S. Boyle and G.E. Smith. With remarkable prescience, Iwama ordered in 1973 that the Centre should 'commercialize the CCD camera in five years for a system price of 100 000 yen' (about \$370 at that time), and with the characteristic Japanese passion for slogans, he declared, 'Our rival is Eastman Kodak.'

Iwama unfortunately did not live quite long enough to see his dreams come to fruition. Sony later honoured his crusading contribution by mounting one of his beautiful sensors on his memorial stone.

The camcorder, as Sony's electronic movie camera is now called, has been an enormous success, and the development costs, which must at one time have brought Sony perilously close to the brink of crisis, now amount to only a tiny fraction of the camcorder's revenues, as Iwama said they would.

It is rarely the case, however, that such long lead times are planned for at the outset. On the social scene, the world is a complicated and uncertain place. Even in relatively recent times we have been rocked by two oil crises, the electronic revolution, glasnost and perestroika with their attendant consequences, and other shocks, none of which was predicted. Each has led, or is leading, to radical changes in our lives, and to the destruction of the careful and rational assessments of the long-term planners and forecasters. But eventually we adjust, and what was once radical becomes so commonplace as to establish a new trend, whence the planners set to work again with apparently undiminished enthusiasm. They will, of course, be determined not to repeat their previous errors in failing to anticipate the last upheaval. And their confidence will be increased by the seemingly endless analyses following in the wake of major change, which often conclude that if we had been more careful and read various writings on the walls we should not have been taken by surprise. The logic of hindsight of this kind can be difficult to resist, especially as there is usually a latter-day Nostradamus somewhere who can claim to have predicted the momentous event, but no one would listen. There is indeed a vast literature, and it is not surprising that some of it might turn out to be accurate.

Science and technology are no less complex, and it is often the case that delays and cost overruns occur as a result of events over which developers have little control. This is especially true, of course, when the technology is new. When new materials like carbon or glass fibres, ceramics, or complex composites are used in a manufacturing process, extensive experience of familiar materials like steel is of little help in predicting the behaviour of the new materials before the composite components are fabricated. Codes of practice to which engineers must conform are specific to each material, and little is known about the ways in which they may be effectively combined. In these circumstances, therefore, each component must actually be constructed, and its performance measured and compared with the specification. If it is not acceptable, the entire process, including the design, the making of new jigs and moulds, and every test, has to be repeated until it is.

Yet, as we saw in Chapter 3, it would be virtually impossible for an academic researcher to get substantial support for research aimed at deriving a theoretical framework for the analysis of new materials unless it was for a particular application. In the absence of such a general theory, manufacturers have no alternative but to resort to expensive and lengthy trial and error.

Technology has usually emerged as a result of a symbiotic combination of academic and industrial research, but it has traditionally been an academic responsibility to map out the broad foundations of understanding on which technology might be built. Some basic research has also been carried out in industry, but industry today is cutting back on research that is not aimed at improving the bottom line, and the academic sector is progressively being pushed into a more mission-oriented role. Unfortunately, the more the foundations are neglected, the higher will be the probability that a technological development will stray into uncharted territory, and lead times will be extended.

There is another reason why protracted projects are likely to be strangled at birth. This is because the champions of a new technological venture will have to persuade a company board that the investment would be justified before the project can be started. At that crucial stage, questions of scientific viability will probably not be paramount; consideration will probably focus solely on the chances of financial success.

If a company invited you to invest, say £1000 a year for five years with the prospect of an income of £1500 in each of the subsequent five years, would it be a good investment? On straight cash-flow terms, the answer seems obvious, for you would show a profit of £2500 in 10 years' time. But if you were to discover that a nearby bank was offering 10 per cent per annum interest for long-term investments, you might change your mind once you realized after a slightly more tedious calculation that putting your money in the bank would yield a profit of more than £4700 after 10 years.

A company board must, of course, consider all the possible uses of the resources at its disposal as part of its duty to safeguard the interests of its shareholders, and the champion of any product or new initiative, technological or otherwise, must be able to show that the proposed investment compares reasonably well with the alternatives. At that stage, however, the monster of 'discounted cash flow' usually raises its hideous heads, and the champion of a project will need to justify that title if he or she is to prevent the proposed product from being cruelly cast into oblivion.

The battle with the monster will go something like this. Its first head will want to know what the so-called net present value of the investment might be. This quantity automatically takes into consideration the levels of investment required, the lead time to market, and the projected levels of income over a certain period, say ten years. The net present value is also what the champion's wonderful project will be reduced to when the monster has finished mangling it. It wreaks its havoc as follows. Imagine you are driving across a flat plain towards a mountain range many miles away. At first, the mountains on the horizon will seem dwarfed by the distance, and it would be difficult to convince someone who has never been there that they are truly magnificent, and among the most impressive sights one can see. This is because Nature discounts, so to speak, size with distance, and the apparent size of an object will halve with each successive doubling of its separation from you.

Discounted cash flow treats time in a similar way. Suppose a champion were to offer the attractive prospect of an income of £1 million a year for five years, if the company were to invest £250 000 a year for ten years until those happy days arrive. It might seem a good thing to do, but the monster will soon chop the vision down to size. At a discount rate of, say, 10 per cent per year, the promised total income of £5 million in the distant future reduces, after another calculation, to a value of just over £1.6 million today. That is, the monster would argue, it would cost only £1.6 million today to buy an annuity of £1 million for five years starting in ten years' time if the interest rate were 10 per cent. But the company's outlay would be nearly £1.7 million in today's money, and so the net present value is therefore negative. The champion might argue that it is only just in the red, but another head of the monster might then point out that the 10 per cent offered by a bank is virtually guaranteed and risk-free, whereas the promised product might not work. A higher discount rate should therefore be applied, whence the putative project would be pushed deeper into deficit. Even if this argument can be resisted, yet another head of the monster might accuse the champion of over-estimating the projected income; and others might also pop up to claim that the estimate of ten years for a successful development is too optimistic, or raise problems about inflation. It is hardly surprising therefore that companies, or government ministers at the national level, can only launch major new initiatives to the extent that the monster of discounted cash flow can be muzzled or reined back.

Unfortunately, these financial monsters, like the Rottweilers and pit bull terriers of the canine world, have their dedicated following, for they give finance directors and others the power to curb a champion's

plans without having to struggle to understand the projected technology. From the scientists' point of view, the monster stifles innovation.
The company board, and particularly the chairman, will therefore
have to weigh the competing claims, and perhaps introduce other
considerations, such as how optimistic or ambitious they are about the
company's technological future.

One of the problems is that only those blessed with hindsight can
foresee how a genuinely new technology will eventually perform. How
could Kazuo Iwama for the camcorder; or James Black for 'Tagamet',
the drug produced at the Welwyn Research Institute in England for
Smith, Klein and French in the 1960s and 70s that inhibits the flow of
stomach acid and cures ulcers without surgery, which took fourteen
years to develop; or John Bardeen, Walter Brattain, and William
Shockley for the transistor; and hosts of others *know* that their work
would have successes that would have astounded even the most starry-
eyed champion, and provided weapons that would have vanquished
the most voracious monsters?

Few turn out to be as successful as these, but the probability that
new proposals will survive a board's scrutiny will depend on its levels
of technological literacy. The chances of success for a project should be
no more difficult to assess than questions on exchange rates, inflation,
fluidity, debt–equity ratios, and the like, that also contribute to an
organization's financial health, questions to which most boards give
frequent in-depth consideration. Company board discussions merely
reflect the expertise of their membership. But intellect is indivisible,
and the executive with penetrating vision in one field should have little
difficulty cutting through the confusion in another. Scientific issues are
conceptually no more complex than many others but, alas, in many
parts of the world, they are rarely discussed with rigour; many boards
rather seem to exude as much enthusiasm for scientific matters as
children for bedtime.

The financial risks of new technology should not be ignored, of
course. But a company also takes risks with its competitiveness and
hence its survival when new technological opportunities are turned
down. Thus, some companies might prefer to hedge their bets by
waiting, and watching how other more adventurous companies get on
before committing themselves.

Unfortunately, however, if every company consistently applies the
same financial rules without rebate, there will be no pioneering companies. We have seen in the academic sector how difficult it is for
researchers to convince committees that they should be allowed to

follow radically new lines of enquiry, and how as a result the subjects of academic research have now become virtually the same in every advanced country. Imagine, therefore, the difficulties that will be encountered by someone who tries to convince a board to put their own reputations and resources on the line when other companies are sitting on the fence. And the more that companies avoid new technology, the narrower will be the niches they can occupy, the more difficult it will be to maintain competitiveness, and the easier it will be for new competitors to enter the market-place as the old technology becomes more widely understood.

These difficulties become compounded during a recession. Indeed, it could be the case that some of the problems of the early 1990s were the first symptoms of a loss of diversity as customs and practices become harmonized throughout the world. But there is nothing inevitable about all this, and only the tiniest of changes is necessary to restore the scientific enterprise to its formerly healthy state.

In the natural world, it has long been known that the health and development of all living organisms depends crucially on regular uptake of trace elements and vitamins. The quantities required are minute. If we allow, for example, our daily intake of vitamin C (ascorbic acid) significantly and consistently to fall below a score or so parts per million of our normal food intake, even though it may otherwise be adequate, we shall probably become ill; and if the shortfall is severe or prolonged, we shall probably die, as many an ancient mariner and others have painfully discovered. Restoration of the shortfall will, however, almost certainly lead to a complete recovery at any stage of the deprivation, perhaps even up to the point of death.

Our society and institutions, too, it would seem, are dependent on a similar requirement in that a small amount of change or renewal or expansion of interests seems to be essential. Nowadays those that over an extended period opt for 'stick-to-our-knitting' policies, or worse still contraction, rarely seem to survive—as, for example, the British motorcycle industry among others has found to its cost.

A small amount of change is like a breath of fresh air, the spice of life, or a shot in the arm, all of which can be nasty taken in excess. Stagnation is suffocation. It must have been miserable living in Western Europe in the Middle Ages. Almost everyone had an allotted place, and there was little that one could do about it. Work was largely drudgery that was repeated generation after generation, and the word 'change' might mostly have been used to describe the stuff that jingled in the purses of the rich.

Then, in the span of a human lifetime or two, a few hundred rebels began to breathe life into this stifled society. Their ideas spearheaded a cascade of change that has persisted, off and on, to the present day. Numerically they were few, but the changes they initiated were dramatic, and affected every material and intellectual aspect of life. Today, however, almost every governmental, industrial, and institutional decision tends to discriminate against such people, and thereby systematically blunts the spearhead of advance.

If these remarks have found a receptive reader, it will have been noticed that the spear can readily be sharpened again with just a little effort. Inspirational people need very little encouragement, but they should not be expected to conform to an agenda agreed by the consensus of others. Nor can their priceless contribution be accurately measured in advance. As we have seen, these individuals can transform the behaviour of those around them, and can be as essential to an interesting life as vitamin C, or even oxygen itself.

Projects can be inspirational too. In one of his first acts as President of the United States, John F. Kennedy in May 1961 committed his country to the *Apollo* programme which entailed sending a manned spacecraft to the Moon, 'before this decade is out'. Extended studies led to the selection of a plan by which a *Saturn V* rocket would launch a 50-ton spacecraft into lunar orbit. The spacecraft contained a lunar module that was to land softly on the Moon, and later to launch and rendezvous with the mother craft. After intensive preparation, and a tragic accident that killed three astronauts, the deadline was met, and on 20 July 1969 *Apollo 11* containing Neil Armstrong, Edwin Aldrin, and Michael Collins was orbiting the Moon. The descent was made shortly afterwards by Armstrong and Aldrin. Armstrong was the first to leave the Module with the now-famous words 'That's one small step for a man, one giant leap for mankind' as he stepped into the powdery surface of the Mare Tranquillitatis, one of the seas (maria) that make up the familiar face of the man-in-the-moon. He was joined shortly in his moonly romp by Aldrin, and thanks to a TV camera they had thoughtfully installed some distance away, Collins watched with mother, and the world watched with wonder.

I witnessed all this at the Frascati Electron Synchrotron Laboratory near Rome in Italy, where I was collaborating with a team of Italian physicists. Hundreds of people crowded around a TV set in the canteen lounge laughing, clapping, and cheering as the men played and joked on our nearest heavenly neighbour, a scene that was seen by many hundreds of millions of people throughout the world.

The *Apollo* programme, however, had many critics both before and after its dazzling culmination, and as other flights followed. What had we learned that we did not already know? What could we do as a result of the multibillion-dollar development that could not be done before except, as some extremists jibed, make non-stick frying pans? Many saw the programme as merely another move in international one-upmanship, in which the US had upstaged the erstwhile Soviet Union's feat in putting Yuri Gagarin into an Earth orbit on 12 April 1961.

The question of whether the programme was the *best* use of the United States' resources is perhaps impossible to answer, but all the criticism ignored *Apollo*'s inspirational impact, as countless millions, and young people in particular, had their imaginations fired by the dramatic extension of mankind's horizons and the sheer audacity of acts carried out in the full glare of global publicity. And Neil Armstrong's famous line might well qualify as one of the quotes of the century.

In Europe likewise, it is commonly held that *Concorde* was one of the great governmental follies and a great waste of money. The Anglo-French project to produce a supersonic passenger aircraft was launched in 1962. After the expenditure of a billion pounds sterling, it went into regular passenger service 14 years later, on 21 January 1976. *Concorde* transports 100 passengers 3500 nautical miles at just over twice the speed of sound for some 40 per cent more than the normal subsonic first-class fare. By today's standards it is noisy, and it produces higher levels of other pollution than would now be acceptable, but standards have of course changed over the years.

Even though its technology is now some thirty years old, *Concorde* still commands attention wherever it goes. Recently, on a flight in a Boeing 747–400, itself an aircraft of considerable achievement, the captain in subconscious tribute drew our attention to a *Concorde* flying nearby, no doubt so that we could all admire its grace and beauty. Its promotional trips still draw large crowds to see it, perhaps to go inside it, or better still, to fly in it. It has acted as an ambassador for Anglo-French engineering in most countries of the world, and it has probably inspired countless youngsters and others to want to fly it, to design something like it, or otherwise do something exceptional themselves. Such inspirational projects exude excitement, flair, and panache, and relieve what can otherwise be tedious and repetitive lives. Yet such intangible factors are rarely taken into account in the balance sheets.

It is difficult for individual companies today to justify their un-assisted involvement in such projects. Their calculations are necessarily more prosaic: unless new projects generate at least as much tangible money as they cost, the company eventually goes out of business. Nevertheless, an inspired or interested workforce will generally be more productive and loyal to their company than one that is worn down by constant calls for penny-pinching cost-cutting.

On the other hand, politicians understandably fear that excitement equals expense, and that their precious budgets and future career prospects might disappear into bottomless pits if they courageously ignore the cautious advice of their counsellors. They receive innumerable petitions of course, all of which cannot be vital, and inevitably they will face the dilemma so eloquently described by Frank Press, namely that political leadership has no way of gauging the volume of resources necessary to maintain the strength of a nation's science and technology.

Rich and altruistic individuals and private trusts and foundations can help, but they are few, and their finance and tax advisers are no less cautious than their corporate colleagues. Moreover, as was mentioned in Chapter 5, private organizations use virtually the same selection procedures—peer preview, consensus, etc.—as the national agencies, and private individuals understandably surround themselves with people whose job it is to protect them from exposure to nonsensical ideas.

For better or worse, therefore, the care of our intellectual well-being in these respects generally resides with government. Unfortunately, good government is increasingly being equated with sound fiscal policies, including cost efficiency and the careful management of the taxpayers' and therefore the voters' money. All very sensible if it can be achieved, but it leaves little room for those who would stir the phagocytes and keep us on our toes.

Sensible it may be, but those at the tip of the spearhead rarely make sense in the beginning. The appreciation of what makes sense is based on experience and education, or in the words of H.G. Wells '. . . the lightly held beliefs and prejudices that came down to them from the past'. George Stephenson's ideas for passenger transport by rail were judged to be 'trash and confusion' by his influential critics; Albert Einstein's ideas were described by *The Times* as 'an affront to common sense'; Langmuir's work on the electric light bulb defied convention; initially there was no demand for Akio Morita's Sony Walkman until he created it; and so on. In my own experience, the proposals of almost every Venture Researcher have been roundly rejected by his or her peers. For people like Stephenson, dissidence is their delight, and the more cautious the selection rules the more certainly innovation will be strangled, and we shall move towards stagnation.

So what is to be done if we are to avoid this depressing fate? National governments, funding agencies, and others are constrained by consensus, and industry seems unable to lift its head from the bottom line. Consider the following, as they say in examinations.

Frank Press's astute observation is the key. It should be remembered, though, that government responsibilities in these respects are of relatively recent origin. Little more than a century or so ago research was funded by individuals and by subscription. In 1854, for example, Lord Harrowby, giving his presidential address to the British Association for the Advancement of Science meeting in Liverpool, said that 'the magnificent State of Britain' had done little for the science upon which her wealth, power, and very existence depended. Indeed it was not until the 1870s that the British government began to give serious support to science, and to follow an earlier lead given by Germany and France.

The growing requirements of science soon outstripped what could be provided by the altruism and the personal interest of individuals. As science became increasingly successful and its demands increased, responsibility for its nourishment was gradually taken up by governments with an enthusiasm that varied with their appreciation of science's capabilities. Governments function by proxy, of course, and they are responsible to the electorate for maintaining and developing the policies they believe will best bring about economic prosperity and national well-being. Although governments themselves cannot generate wealth, the frameworks they set up will largely determine the success of those who seek to do so.

The sources of economic growth are diverse. They include the banks, stock exchanges, and other financial institutions that provide, among other things, the vital lubrication for a national economy. The volume of their transactions is prodigious, and the same commodities may be bought and sold many times without those involved coming into contact with them. These transactions are the market, of course, and they determine the values of the goods and the companies that make them. Ultimately, however, on the global scale, the agricultural and the manufacturing industries, including the suppliers of power, water, and other utilities, the providers of transport, communications, and other facilities of an efficient industrial infrastructure are the primary determinants of growth, and most of the multitude of financial transactions must eventually terminate on one of its outputs. For industry, one might say, the buck stops here!

In principle, therefore, it might be thought that responsibility for supporting the scientific enterprise should reside with industry: as the generator of growth industry is best placed to resolve Frank Press's dilemma. Each company knows its business and the investment necessary to maintain it in good health—including investment in research. But the scientific enterprise has many outputs including training, education, and culture as well as research; and much research is carried out in support of national requirements, such as the protection of the environment. It therefore seems right to expect that government should provide for the national scientific well-being on our behalf, and, of course, most governments do so.

There are limits, however, to a government's capabilities. When the pace of change is slow and the future can reasonably be planned, the actions of a competent government or any other democratically based organization should be sufficiently effective to satisfy their constituencies, as indeed they are in the overwhelming majority of cases. But major scientific or technological advances cannot usually be planned. In the 1960s, for example, who could have commissioned the generic discoveries that led to such spectacular successes as transistors and integrated circuitry, the jet engine, photocopiers, opto-electronics, polymers and plastics, and others that have contributed enormously to economic growth? There seems to be a dearth of such new advances today, and some industrialists are expressing their concern at the lack of new opportunities for growth.

I believe that industry and other sources of private wealth have a key role to play in alleviating this serious situation. Financially speaking, it should not be too demanding a responsibility since after all, the number of people capable of precipitating radical change in science and technology is very small, and it would seem reasonable that industry should take a lead in their encouragement and, thereby, in restoring a healthy diversity to the enterprise. One of the difficulties here is that industry has long been averse to supporting genuinely exploratory basic research because the outcome is unpredictable; and since any one company usually has a relatively narrow range of interests, the benefit might easily fall outside it.

Imagine, however, a new type of industrial agency set up explicitly to catalyse change in ways that combine altruism with good business practice. Imagine a consortium of companies that in the main do not compete with each other, but when taken together have a collective range of interests that spans the entire industrial spectrum. Imagine these companies protectively encircling, like a ring of covered wagons, a small collection of groups of researchers, each on a mission to explore areas well outside the current priority fields, and to improve understanding where it is weak or non-existent. Any discovery to come from the research would be of potential interest to one or more of the industrial 'wagons'. Furthermore, each company would have privileged access to any patents, and when there were successful developments royalties would be shared between all the companies, the researchers, and their institutions according to a previously agreed formula. We would then have an arrangement whereby every participant in the initiative would gain. The researchers would thus have the chance of showing that their radical view of the world was justified. We know from more than ten years' practical experience of a similar type of agency (which in our case is known as Venture Research for want of a better name), that there is a high probability—more than 50 per cent—of potentially significant industrial opportunity arising even in the short term of a year or so; and there is always the prospect in the longer term that the work could be of major importance and could lead to completely new types of industrial opportunity.

The intrinsic cost of research of this type is relatively low because the scientists involved are not in competition with anyone; and when one enters a new and untilled field even modest equipment can yield an interesting crop. In addition, the number of scientists and engineers with the courage, ability, and the inclination to embark on pioneering crusades is very small, perhaps only a few per million of population in an advanced country, and so overall costs too will be relatively low without having to impose arbitrary and restrictive priorities or budgets.

So much for the overall structure of this business-led quest to restore diversity to the scientific enterprise. But how might these pioneering researchers be selected, and how might their inspirational ideas be made to pay?

Nowadays the ubiquitous peer-review system reigns supreme, and it is widely believed that there is no practicable alternative for the assessment of basic research. Nevertheless it is peer *preview*, of course, and it works well enough for mainstream research as we have seen. But how can pioneers get permission radically to challenge the status quo their peers generally accept when the evidence does not yet exist that it might be flawed? It is, of course, not completely impossible, but it presents a layer of difficulty that hardly existed a few decades ago. As has been mentioned earlier, researchers were then free to do whatever they wished with what little money they could raise. It was never enough, of course. But it was not so much a cause for concern as a fact of life. The problems arise when scientists are forced into competition and set out to do so-called 'world-class' research. Money then becomes crucial. One cannot expect to compete in, say, Formula One motor racing without access to the latest and best equipment. When one has the best, then by conventional definition, one's work will be world-class. It is not the only way forward, however, and the world-class use of the equipment we all have between our ears can lead to astonishing outcomes. Freedom is the key, and the Venture Research approach to re-sharpening the spearhead is to restore that freedom to those scientists and engineers who can demonstrate that they need it, and want to do something genuinely new.

Venture Research selection procedures can work only if they are operated by people who consider themselves as practising scientists rather than administrators or managers. That is, they are scientists prepared to exchange their hands-on involvement in research with a role that entails taking vicarious pleasure in other people's work in return for the vast panoramic range of interactions they will thereby enjoy. Selection is almost entirely at the conceptual level, and we look for applicants who can demonstrate strategic vision in their research. For the vast majority of scientists it is usually the case that their tactical expertise can be assumed: that is, they are fully competent in the uses and limitations of the techniques to be used in their research. Peer preview, by the way, focuses almost entirely on tactical considerations, which is hardly surprising since the peers who do the previewing are usually drawn from the ranks of those who accept the general concepts on which the proposed research will build. Venture Research selection, on the other hand, proceeds much like a discussion on novel approaches to military strategy between people from different forces—the army and navy for instance. Although the nitty-gritty problems each force has to face are vastly different, neither would expect to have difficulty in recognizing strategic flair and panache in the other.

Venture Research proposals are unsolicited, for otherwise we would approach only the famous and the well-connected. *Anyone, anywhere, of any status* may apply. This means, however, that we have to take responsibility for spreading the word that we are looking for challenging proposals, and paradoxically, that is not easy. The main problem is that researchers do not expect to hear about an initiative coming from the industrial sector that is apparently more adventurous and altruistic than the national funding agencies. The advice from the hundreds of scientists we consulted was that we should not advertise in the usual way, since it would be virtually impossible to convey the subtleties of the Venture Research message in an advertisement. Since this advice confirmed our own prejudices, our conclusion was that we in Venture Research should broadcast our message personally 'on the hoof' through talks and meetings, etc.; and to seek every opportunity to describe what we were trying to do in every journal, newspaper, radio, or TV programme that would allow us to do so. This second course proved especially difficult, for the media suspected that we were trying to get cheap corporate advertising for our patron BP, even though we were always willing to agree that BP's name need not be mentioned.

Pioneers must persevere, of course, and we usually manage to find a way round these problems. Our invitation is for researchers to send short proposals, typically a page or less, or to make a phone call, briefly describing the research. Not surprisingly, the great majority of proposals, about 80–90 per cent, turn out to be requests for funds to extend previous work, or for development, which we either decline after checking that our impressions are accurate, or, with the originators' permission, pass to others who might be interested. Leaving aside the inevitable few that claim to have discovered perpetual motion, we usually find that about 10 per cent of our harvest strikes a spark. These people are invited to a meeting where their proposals are discussed at length in face-to-face dialogue. It is curious that virtually every other funding agency selects on the basis of *written* proposals only, although writing is so difficult, and academics are no more skilled in this art than others. Speaking, of course, is an academic speciality. Why not let them show it off?

The discussions centre on such questions as what do you want to do? Why is it important? How does it compare with other work done elsewhere? Where are the departures from orthodoxy? What would be the consequences if your novel view of the world were shown to be justified? We do not have a check-list, but our main purpose is to try to understand what motivates the scientists or engineers who come forward (alas, there are few engineers), what is their vision, and why they are dissatisfied. We try to share some of the excitement of their work. We must also win their respect if our discussions are to have any depth, not only during selection, but thereafter. This is perhaps the most difficult part of our *modus operandi*, for it is so easy to misconstrue what is said and fail to distinguish it from the flattery that is often offered to possible sources of money.

The individuals we are looking for are extremely rare, and so it is inevitable that we accept very few of the proposals put to us. Consequently we have to decline the great majority, but we always make it

clear that we are expressing an opinion on the degree to which a proposal satisfies our well-publicized criteria. We are not so much reject-
ing people as inviting them to respond with something more exciting
and potentially revolutionary. Of those we invite for discussions, it is
remarkable how many scientists warm to this type of meeting and say
how much they have enjoyed the interaction, even though they will
usually have gained nothing tangible from it. Many have said that they
have never had to respond before to questions of the type we ask, and
they have been grateful for the ways in which we help them work
towards the answers.

For the rare few, however, it is even more remarkable how they
select themselves, as, of course, they did in the past. We merely create
the environment that allows them to do so. We must take care to try
not to be seduced by fallacious reasoning, and to make allowances for
variations in loquacity. One of our Venture Researchers stammered,
as I once did myself, but I suggested that his stammer might go when
his new work became successful. It did. Others who came forward
could charismatically charm the monkeys from the trees. For our part,
it can often be very difficult to absorb the gist of an argument, but the
reward when the daylight dawns is like making a scientific discovery
oneself. To see, at last, in a sudden flash, the point of a discussion, and
to recognize the powerful significance of a completely new way of
thinking is one of the greatest pleasures. We realize, however, that it is
inevitable that we shall occasionally be mistaken in our selection, or the
scientists themselves may be unable to climb their Everests, or find that
they are unclimbable. These are risks that we must be prepared to
take.

Once we have found our Venture Researchers, they are given the
resources they require and every encouragement to follow their
research wherever it may take them. How then do we extract value
from this apparently uncontrolled situation? Reams have been written
on technology transfer, and countless committees have been asked to
advise on ways to improve the efficiencies of transfer between the
academic sector and industry, which is often low even when the
research has been commissioned.

From our point of view, transfer is usually attempted far too late.
Typically an academic may have developed an idea, or process, or
entity that is believed to have value to an industry. The invention may
work well enough in the academic environment, but there is a vast gulf
between a university laboratory and the market-place that must be
bridged if the invention is to have value. The academic must therefore

seek out an industrialist, but the academic may be afraid to reveal all, because even if the invention is protected by patent, there is a belief that a powerful industry may have little difficulty in circumventing it, or in finding some way of reducing or avoiding payment of the rewards that the academic believes are justly deserved. A dance of the seven veils often ensues simply because there is no mutual trust.

Mutual trust is, of course, essential to all cooperation. Without it, there will probably be breakdown, sooner or later. One might ask a friend to do something reasonable and be confident that he or she will oblige; a stranger will almost certainly not do so. But it is very difficult to allow one's trust in a person or organization to grow at the *same time* that one is trying to protect one's interests.

One of our responsibilities, therefore, is to establish trusting relationships from the outset, but particularly *before* they are put to any test. Luckily, this is not so difficult because most, if not all, people would usually prefer to be friendly if their interests are not threatened. Initially, we arrange informal gatherings between Venture Researchers and industrialists that concentrate on discussing the general potential of the research, so that when tangible opportunities eventually emerge, as they almost invariably do, minds have been prepared. 'Not invented here' problems do not usually arise in these circumstances because whatever emerges *has* been invented here. Everyone concerned will have had a part in the outcome, and everyone stands to gain if it can be shown to be tangibly successful.

In practice, it works as follows. In the early 1980s British Petroleum had extensive research and development capabilities in every sector in which the company had interests, except one. Although BP is a major user of computers and computer programs, there was no forum within BP for the discussion of new computing science or for the assessment of its commercial potential. BP therefore had little alternative but to accept what the computer manufacturers and computer software market could provide. This situation would have been unthinkable in any other field, and it arose simply because BP was not aware that other courses of action were open. In the natural sciences, the procedures for the evaluation of discoveries, for the identification of specific applications, and for their development into reliable and economic products for sale in the market-place have evolved over many decades, whereas the equivalent procedures for computing science were then still in their infancy even for specialist computing companies. In BP they did not exist at all.

In 1985, however, following an extended series of Venture Research workshops (affectionately described as 'love-ins' by one senior BP executive) and some gentle lobbying, BP set up an Information Technology Research Unit that progressively transformed BP's prospects in this vital field.

Nowadays there are new research initiatives in profusion, but it is rare indeed if it is not argued that the probability of their success will be directly proportional to the size of the investment in them. Unfortunately, politicians and the captains of industry who carry the burden of evaluations have little to guide them, as Frank Press pointed out, and see only the prospect of bottomless pits.

The limits on costs of the type of agency I have in mind are set, however, by a much higher authority, that is by the limitations on creativity itself. Thus, in the case of Venture Research, after ten years of unconstrained growth during the 1980s when it was patronized by British Petroleum, the total budget that covered research in Europe and North America was some £2.8 million per year, even though much more money was available. This budget also included the industrial overhead of basing the agency's core staff (the Venture Research Unit) in the most expensive part of the City of London. Research was supported in whatever location the Venture Researchers chose, which, of course, changed as careers developed.

This pilot scheme was very successful in business terms, as can be judged from its survival and expansion during ten turbulent years in an industrial environment, with all its attendant slings and arrows of outrageous fortune. The scheme attracted particular attention in the British academic sector and helped to enhance British Petroleum's image and reputation for association with excellence, which in turn helped graduate recruitment.

Tangible success is, however, more difficult to demonstrate. British Petroleum is a conservative company. Its interests are relatively narrow and cover only a tiny fraction of the 'ring of covered wagons' we have in mind. In addition, the political thicket is no less dense in the scientific enterprise than elsewhere, and there were those who saw Venture Research as a threat rather than an opportunity, perhaps because its bang-per-buck was so high. Perhaps one example might be sufficient to indicate how frustrations can arise. At a meeting in Tucson Arizona in December 1987 to celebrate the 75th anniversary of the Research Corporation of America, I met Professor Lewis Branscomb, and mentioned proudly that Dudley Herschbach had just joined

our community of Venture Researchers. Branscomb was then Director
of the Science, Technology, and Public Policy Program at Harvard
University, and had recently retired as Chief Scientist at IBM. He was
delighted, and very congratulatory. 'Do you realize', he said, 'that you
should have a million dollars finder's fee? Herschbach is inspirational
and will have a big impact on your company's chemical operations.'
Unfortunately and sadly, despite repeated attempts to do so, we were
not successful in setting up even one working meeting between Herschbach and BP's industrial scientists, though he and many of our scientific colleagues were more than willing to take part.

Luddism is far from dead, of course, and probably will never die.
The Venture Research initiative must eventually be able to cope with
this and any other type of opposition, of course, and will not be viable if
it cannot. Its advantage lies in its low cost and its association with
virgin fields rather than bottomless pits, and its potential for leverage is
consequently high. Most important of all, however, is the possibility
that the initiative begins to provide a response to Press's dilemma
introduced on p. 120.

The achievement of economic growth seems to be essential to a
nation's health, and every recession brings a grim harvest of misery
and misfortune. It now seems to be generally accepted that technological change is an essential ingredient for growth, and that change
can be better accomplished with a sharper spear of scientific advance.
By forming consortia or other means, companies can ensure that the
spear is kept as sharp as possible, while at the same time retaining their
control of the extent to which the new science is translated into new
technology. New science, like money in the bank, does not always have
to be used to the full. Rather than live a hand-to-mouth existence, or
worse, to borrow at high rates of interest to finance excessively long
lead times, companies can ensure that the balances of scientific reserves
are as far in the black as human ingenuity can provide, while each
cooperating company would also be acting in its own interests.

In this model, therefore, the determinators of growth would become
the patrons of scientific leadership. Although each patron would have
the first option on all technological potential that emerges, most if not
all the underlying science would pass into the public domain. Again,
this apparent altruism would be in the interests of company–patrons,
since the new science provides the inspiration for tomorrow's scientists
and the next technological crop.

Scientific leadership can be inspirational among a much wider community than industry, and governments could respond by subsidizing

this type of activity and encouraging the participat
Some governments, the Japanese for example, do not t
capital gains, and companies and their investors are enco
plan for growth.

A response to each nation's Dr Press might therefore be that politi
leadership should no more be expected to gauge the amount of resources necessary to maintain the strength of a nation's science and
technology than they are to determine the strength of a nation's industrial activity. Governments will usually strive to create the frameworks that maximize the latter, but it is men and women in industry
and other institutions who together carry the responsibility for the
levels of activity that will in fact be achieved. A government's proxy
function on behalf of a population, and its accountability to it for the
exercise of that function, works well enough for resource allocations to
such sectors as hospitals and highways, and indeed for any subsidies to
industry too, but it cannot be expected to provide scientific leadership,
since Nature does not respect democratic methods or opinions.

The changes required to foster the restoration of leadership to those
few individual scientists with the capacity for it are small, but existing
infrastructures nevertheless do not seem to be able to accommodate
them. We are suggesting here that industry should take responsibility
for ensuring that the scientific enterprise enjoys a healthier and more
diverse diet. The Venture Research initiative is one way in which
industrialists can combine their ingenuity with the new insights offered
by genuinely pioneering scientists. Other ways will emerge once it
becomes generally accepted that a dash of dissidence is an essential
ingredient for the achievement of growth.

8

and the future

ınd the five billion or so other folk that share this pla....., p........rs, is something of a Johnny-come-lately. The Earth has been a going concern for more than four billion years, whereas the species we recognize as ourselves crept on to the scene only about half a million years ago. These huge numbers are, however, hardly user-friendly, and it may help at this stage to put on a pair of 'chronacles', which by analogy with the 'mathacles' I created in Chapter 3, trim the time dimension to more easily appreciated proportions. If we choose those 'chronacles' that telescope the Earth's real age to one year, we shall see that we have been around in our present form for barely an hour. Mankind's entrance was also unobtrusive, and for most of this time left little behind in the way of permanent record apart from such relics as a few fossilized footprints, the occasional collection of caveman's cartoons, and some overgrown barrows of buried bones.

For the past ten millennia, or about a minute through the eye of our 'chronacles', mankind's innate and irrepressible industry has been increasingly making its mark. Nowadays, many human endeavours have a global dimension. Industrial and other effluents and emissions can affect the air we breathe, the rivers, seas, oceans, and perhaps climate. Demand for scarce resources such as energy, water, and food can cause international conflict. Nuclear weapons have the power to threaten the planet's ability to sustain life. The pressure of population is a threat to stability in many parts of the world. In the less developed countries, to use the current politically correct euphemism, disease is still a major killer, while in the developed world the persistence of cancer and other potentially terminal conditions, that have generally withstood the subsequent onslaught of many billions of research dollars, pounds, marks, francs, etc., and more recently the emergence of AIDS, have tended to give support to the idea that we have reached a watershed in our development.

There is no question that mankind's emergence from a brutish existence, and primordial parity with the other primates, has been brought about by the marriage between intellect and technology. As we have seen, the relationship grew out of an inspired improvization that was later to be transformed by a relatively few hungry scientific souls, who not only made possible a vast range of improvements to the old ways but also brought forward the potential of new technologies that no one had dreamed about. It is hardly surprising, therefore, that some people hold scientists to blame for all the prevailing gloom, uncertainty and doubt that beset us. Indeed, in the past decade or two we have seen the gradual growth of an environmental evangelism that is reminiscent of the puritanism of a few centuries ago and is supported by those who, like their post-Renaissance forebears, would like to see a return to stability and to the supposed pleasures of a more pastoral age.

There is also no question that the world is astonishingly complex, perhaps terrifyingly so for some of us, and that despite our impressive progress we still understand very little. Complexity does seem to be infinite in all directions, as we discussed earlier. It can be tempting to dismiss some of it as irrelevant to our everyday problems: the structure of sub-nuclear matter, say, or the origins of the universe and what seems to be our complete ignorance of the form taken by some 90 per cent of the matter contained within it. There are nevertheless complexities that we cannot avoid or afford to neglect because they are part of the fabric and furniture of the world we live in, and the stuff of which we and all living organisms are made. To cite only one example, it was widely believed at the beginning of this century that our Sun was a great 'chemical engine', but thanks to the understanding that has come from Ernest Rutherford's seminal work (see Chapter 1), we now know that the processes that drive the stellar object which dominates our lives are nuclear rather than chemical. Closer to home, the energy generated by other nuclear processes is now used to provide some 70 per cent of the electricity used in France, and smaller but substantial proportions in almost every country in the developed world. All this from a researcher who denied that his work had relevance to everyday life!

The environmental evangelists might argue, however, that the world would be a safer and cleaner place if the science that Rutherford and others uncovered had remained obscured. But, in common with most if not all great scientific pioneers, Rutherford sought to increase understanding rather than to solve any specific or narrowly defined problem. New science is synonymous with new understanding, and those who would curtail science would have their success rewarded by the preservation of ignorance. Knowledge does not necessarily bring happiness, of course, as has been known since biblical times, perhaps because the more one understands the easier it is to see more profound problems, and the relative insignificance of our comprehension of them. As learning is hardly ever painless, why bother to take the trouble?

Societies and the individuals who form them must make their own choices, but those who would wish to see a slowing-down of scientific enquiry and therefore of mankind's economic progress, should realize that they are thereby striving to suppress a trait that is apparently genetic, and which distinguishes us from all other species. We have a tiger by the tail. Those hungry souls and pioneers who would shed new light on our general darkness can be successfully repressed only by being incarcerated or eliminated, which was precisely what was done in the centuries before the Renaissance in Europe, and is still the fate of heretics and dissidents in other parts of the world today.

Knowledge does not necessarily bring happiness, but it can increase appreciation and awareness, which, of course, is why tyrants impose censorship. Science and technology have indeed contributed enormously to the spread of culture, and thereby to an increased awareness of ourselves and to our capacity for contemplation and enjoyment; it must be difficult to appreciate anything when one is starving.

We all know, of course, that science can be used for much less noble purposes in peace and war, but malevolent exploitation is not new, and what we see today differs only in scale and scope from what we have always had to endure throughout the millennia since we succeeded to our present sapient state.

At a meeting in 1989 of the Anglo-Japanese High Technology Forum, the Japanese co-chairman, Hoachiro Amaya, pointed out that the growth and influence of science and technology throughout this century could be thought of as having kept roughly in step with the increases in the speed of powered flight, whereas human comprehension had in general progressed like 'a cow walking'. The reduction of this grotesque mismatch is perhaps one of the greatest challenges faced by mankind. For many people comprehension has not progressed at all. Almost a billion of our fellow-men are 'chronically undernourished', which in the language of bureaucracy means that they do not get enough to eat to sustain body weight and light activity: in other words they are starving. At least as many others succeed in escaping from this most appalling of categories only by endless toil and drudgery that has changed little over the centuries.

The potential for improved food production would, however, seem to be enormous. There are about 80 000 plant species that produce some part that is edible, but only 50 are actively cultivated, seven of which—wheat, rice, corn, potatoes, barley, cassava, and sorghum—provide 75 per cent of the world's food supply. Unfortunately, as was reported in *Science* in 1992, 'crop science has been balkanized, with specialists working in their own commodities and attending their own meetings.' But there are moves towards a more coherent strategy in that *Science* was reporting on the First International Crop Science Congress held in Ames, Iowa in July 1992. Among other things, the Congress noted that many neglected crops are tolerant to adverse conditions such as heat, cold, or drought, climates which, of course, tend to be prevalent in the more underprivileged regions.

This is not, however, the place to rehearse in detail the problems of the world: that has already been done elsewhere, and more than once. Discussion and debate will no doubt continue, but the vital role of science and technology in solving these problems cannot be denied. Many of the remedies that are at present being applied are poorly understood and are likely to offer only short-term solutions, which is clearly unsatisfactory.

Ideally, therefore, we should interfere with natural systems such as crops, animals, forests, seas, or ourselves only when we understand the main features of the complex web of material and social relationships that make up each one of them, and also their dependence on each other. Ideally, ignorance should be abolished, and horses should talk. Sometimes the social or economic pressures for expedient actions in the short term can be irresistible, and as in medicine, it is easier to treat a symptom than to cure a disease. If an action is taken simply because it seems to offer a viable way forward for the time being until the system as a whole can be understood, expedience can be excusable and defensible. But the more people are capable of recognizing short-termist policies based on expediency, and are pressing governments, industry, and the institutions for their replacement by serious searches for solutions, the fewer will be the technological mishaps and misfortunes we shall eventually have to endure.

Nevertheless, if science were to be obliged to yield one single unequivocal message, it might be that Nature's complexity appears to be infinite, and that knowledge will therefore always be incomplete. Mankind's transition from obscurity to the acquisition of globally significant capabilities was never smooth, and although I believe that confidence in science is fully justified, we must expect and understand that every advance will contain a flaw, and that we can make progress towards a better world only to the extent that we are prepared to be adventurous.

Amaya's succinctly expressed concern about the progress of human comprehension is, however, still barely recognized in the relatively affluent developed world after more than a century of universal educational programmes. Indeed, our societies today seem to be becoming increasingly preoccupied with safety, certainty, and stability, and do not yet seem to have realized that stability is, paradoxically, a dangerous state.

There does not seem to be a system in the natural world that does not owe its emergence, survival, and growth to instability. The birth of the universe itself some 15 billion years ago (or about four years through our chronacles) seems to have come from the ultimate in spontaneous instabilities in a system, for want of a better word, we can barely comprehend; for how can one describe the entity from which the Big Bang exploded at a time when the fabric of space and time had not yet been woven? Thereafter, for the briefest of moments the newly born universe seems to have been a hot, featureless glop, but as it cooled its serene symmetry was broken by the emergence of the forces that began to form its complex and changing character: the nuclear, the electromagnetic, the weak, and the gravitational; a total of four forces (so far as we know) that seem capable of generating infinite variety. The formation of our galaxy, the Sun, and our planetary home seem to have been seeded by chance fluctuations. Little is know about the processes that led to the Earth being blessed, about a billion years later, with living, self-replicating organisms, or whether these processes are exclusively confined to our planet. Once life had arrived, however, its dazzling diversity seems to be due to Nature's predilection to change spontaneously and apparently at random the set of molecular components that go to make up a gene. Like the numbers of weekly winners in a national lottery, these mutations seem to go at a roughly constant *average* rate, but their precise location is just as unpredictable. In addition, just to add a little more spice to life, the pack of DNA cards donated by each parent playing in the game of sexual reproduction is also gently shuffled before they are dealt to the next generation. Those species that prefer to avoid sex have their rules for cutting the pack before a similar set of shuffles to ensure that the next generation is not simply a prosaically precise reproduction of the previous one.

Nature's passion for molecular instability does not end with DNA. The most versatile improvisors, the proteins, are responsible for carrying out a wide range of functions in all species. The proteins are complex polymers; they are usually large molecules composed of sequences of amino acids. More than a hundred are known, but twenty, the 'magic twenty' as Francis Crick has called them, seem to be universally essential to protein synthesis. Proteins have a similar role to the nucleotides in DNA (see p. 33) in that they specify structure and function. The twenty-'letter' code found in proteins is, of course, much more complicated than the four-'letter' nucleotide version found in DNA. Indeed, until some fifty years ago it was firmly believed that proteins were indeed the molecules that determined life, since they were uniquely capable of specifying the range and diversity of living organisms. The word 'protein' itself is derived from the Greek *proteis*, which means first rank of importance, a status we now know belongs to DNA.

There are many important biological processes, however, that depend on protein instability, and on instabilities in other molecular species too. Scientists wishing to study these processes in the laboratory (*in vitro*) have had to devise procedures for stabilizing the molecules involved. It was while reflecting on this recurrent problem that Colin Self formulated in 1988 a Venture Research programme with the objective of focusing attention on the biological *advantages* of instability rather than dealing with its consequences. Some protein lifetimes, for example, can be very short, but at present it is not possible to set a lower limit to the range of their transience because present methods of measurement are inadequate. It could be the case that an important part of the range of lifetimes is hidden from us simply because we have neither had the techniques to explore it, nor consequently perhaps, the inclination to do so. Since the living organisms and systems we see today are the most successful to emerge from the rigours of some three billion years of evolution, it would be remarkable if this mercurial transience did not have its advantages. Colin Self, who is now at the University of Newcastle, argued persuasively that it would be very interesting to know what they might be.

In contrast with the natural world's affection for instability, we can see in everyday life that, in the advanced countries at least, governments, institutions, and industries, are tending to become increasingly averse to taking actions that differ from the norm, or that the cautious might argue are risky. In consequence, the flamboyant mosaic of a world that was colourful, in terms of the diversity of its customs and practices, is beginning to descend into grey uniformity. For scientific enterprise, it is now almost invariably required that the proponents of new initiatives should be able to show in advance that they offer the *best* value for money, and also that the chances of the objectives being achieved approach the certainty of the graveyard. But if something has not been done before, these requirements are impossible to satisfy. The only way that George Stephenson could show that very expensive railways would eventually offer a better and cheaper form of passenger transport than low-cost stage-coaches was to raise the billions of dollars in today's money and build them. Luckily for him, discounted cash flow had not yet been invented, nor was the stage-coach lobby and all the others he would have had to contend with as well organized as it would be today.

Societies today are also much more litigious than they were, and the belief is now widespread that if something goes wrong, even if it is entirely accidental, an individual, or group, or organization is to blame, and must pay for the consequences. Not so long ago if I had tripped on a loose paving-stone I would probably have been told by a passer-by that I should be more careful where I put my feet; nowadays I might be advised to sue my local council because they had failed in their duty to provide a level footpath. Legislation on product liability has now become widespread, and, of course, it helps to protect us from irresponsible manufacturers and suppliers of professional services. Our increasing preoccupation with law and litigation, however, also create new sources of inertia for innovators to overcome, and provide powerful arguments against change.

What can science and scientists do about all this? Science, as is well known, is a disciplined route to new types of understanding, and the more we understand, and the greater the diversity of our understanding, the better we shall be able to cope with the world's escalating complexity as population increases. But the more that that knowledge is held by only a privileged few, the more it will be regarded with suspicion and mistrust by the rest. Scientists have been aware of this problem for more than a century, but despite their collective good intentions very little progress has been made in correcting the imbalance. Many elder statesmen of science do an excellent job in disseminating news of the great discoveries, but perhaps because there are too few of them, or simply because their Olympian status makes it difficult for men and women in the street to relate their messages to everyday life, the problem not only persists (as Amaya has so eloquently described) but it is getting worse.

Yet at the working level, individual scientists who would wish to devote some of their time to sharing the joys of exploration with the general public risk losing the respect of their colleagues: it is difficult to avoid the implication that someone who wants to spend less time at the laboratory bench is perhaps looking for an easier option. They would also run the risk of falling behind in the race to be first. In these circumstances, therefore, given the pervasive effects of peer preview and the increasing competitiveness of scientific activity, it is easy to see why very few young scientists can afford to be missionaries. To illustrate the problem from another arena of human endeavour, imagine what little following sport would have if we were not able to watch and appreciate the flair and individual creativity of the current stars, and could only read or hear about them in the reports prepared by their managers for a specialized following.

As we have outlined in earlier chapters, there is at least one viable alternative to peer preview. The Venture Research initiative offers one tried and tested possibility that will help to maintain the sharpness of the spear of change in terms of new discoveries and their industrial development. It also has a wide appeal. During the last two years of BP's patronage of Venture Research, we ran a monthly series of lunch-time presentations by active Venture Researchers under the title 'Entertaining Science'. As one of the tiniest components of BP, Venture Research could not command anyone's attendance, but in spite of the competing demands on people's time during that short interlude from work, and the many other attractions in a major city like London, we always managed to fill BP's largest auditorium with a mixture of some two hundred head-office staff and their visitors, even though the

Fig. 6. Cartoon drawn by Chris Masters used to advertise a lecture in the 'Entertaining Science' series by Professor Robin Tucker (University of Lancaster) entitled 'Einstein's Dream'. By courtesy of the British Petroleum Company.

subjects on offer, such as quantum mechanics, plant genetics, organic chemistry, and plate tectonics, are not usually regarded as having popular appeal or humorous content. Clearly, there is a demand for intellectual stimulation.

Amaya's implied challenge is, of course, crucial. There is no doubt that science and technology can improve the material aspects of a human life, but the less the concepts of science are understood the more the less responsible elements of society will be able to get away with such evils as pollution, and short-termist exploitation. It is possible, however, that science's evident *material* successes, and its association in many people's minds with proof and certainty, may create the impression that science is synonymous with cold calculation, and can contribute little to the milk of human kindness.

At a friend's dinner party recently, our host, a successful business man who is generally widely read, commented sadly on my remarks promoting the value of science by saying that he did not want to see 'the death of poetry'. He could not bear the idea of a life spent travelling along 'teutonically clean stainless steel corridors with no mysteries'. My friend is not alone in holding these views, and in 1992 several influential books were published that were severely critical of science's challenge to religious and moral values.

In the United States, Neil Postman said in his book *The surrender of culture to technology*, 'The uncontrolled growth of technology destroys the vital sources of our humanity.' In Britain, Bryan Appleyard in his book *Understanding the present: science and the soul of modern man* said that 'Science . . . is spiritually corrosive.' After a substantial review of the book, prominently displayed, *The Times* announced that it was sponsoring a debate in London, in collaboration with Dillons and Pan Macmillan, on the motion 'The heartless truths of science strip man of his spiritual dignity'.

Fay Weldon, the British writer, supported the motion, and was reported by *The Times* as saying that scientists 'were robbing us of the hope of heaven—a world without belief is dreary, pointless, and frightening'. Professor Lewis Wolpert from University College London opposed and was reported as saying that science was a tough discipline, not easy to understand, and many people felt excluded by their ignorance of it. But it was wrong to blame science for the secularization of society as Fay Weldon had done. There were many other causes of that. Bryan Appleyard was reported as saying that science's success had led to the assumption that it was the only worthwhile form of truth. The debate was chaired by Melvyn Bragg, who is also a writer. It was very well attended, and perhaps not surprisingly the tongue-in-cheek motion was declared to be 'overwhelmingly defeated'.

It would seem, however, that these adverse views of science are further expressions of the problem of comprehension to which Amaya drew our attention. Science rests ultimately on a set of assumptions, usually called axioms. These are statements, such as that the universe is infinite in all directions; or that science studied on the planet Earth will yield exactly the same results in the same circumstances in any other part of the universe. The truth of axioms like these is impossible to prove, but one of the objectives of science is, or should be, to keep them under continuous review. The concept of time, for example, was implicitly thought until this century to be the same for any observer, anywhere, but Einstein showed that time was strictly a local matter. The axiom of constant time was shattered as soon as Einstein's theory had been accepted. Thus one scientist's axiom might be another's challenge, and axioms are only accepted as long as they survive.

The axioms provide the foundations of a science and convenient points of reference, but they are not tablets of stone, and every scientist has the right to challenge them. Scientists also have a duty to prove to their colleagues that their observations are correct within the limitations of measurement, or that their hypotheses give a better account of a set of observations than all previous hypotheses, or that otherwise their view of the world is justified. The more general concept of truth, however, does not fit comfortably within scientific methodology. The concept of truth as used in everyday speech refers not only to accuracy, but also to such abstractions as loyalty and sincerity, and is intimately connected with the quality of human relationships. Furthermore, if one accepts the axiom that the world's complexity is infinite, the scientific understanding we have at present will inevitably be incomplete, and merely the best we can manage for the time being, whatever our intentions; it will probably be only a matter of time, therefore, before that understanding gives way to a more general view. Scientists who do not accept that or any other axiom have a duty to show why they do not.

In the past, science and religion have been seen as in conflict, and many people today would share that view. My own opinion is that the two are almost completely independent of each other. Be that as it may, I would argue that the discoveries made by scientists, whether those of Galileo, Kepler, or Newton in the seventeeth century or of Darwin in the nineteenth or Einstein in the twentieth, have been spiritually uplifting and enriching. Almost all the major discoveries in science have been made by people driven in Einstein's splendid phrase by 'a hunger of the soul'; many have had little interest in the material success their work has brought, and many have been deeply religious in the conventional sense.

Few people in the advanced countries would believe that increased material wealth necessarily leads to greater happiness or spiritual well-being, and so it is inevitable, therefore, that the more materially wealthy we become the more there will be opportunity for disappointment as the levels of our contentment do not automatically keep in step with our riches. Since economic growth has been almost entirely attributable to developments in science and technology, it is hardly surprising therefore that some people will make science a scapegoat for their spiritual malaise.

It is a pity that educational programmes place so much stress on the role of science as a generator of technology, and in turn that so much technology is devoted to providing forgettable froth for our free time. To make matters worse, the constant search for economies of scale in the workplace has led to the development of user-friendly technologies that are intended to rely as little as possible on the quality of human performance, and so it is becoming increasingly difficult to find intellectual stimulation and satisfaction there too.

Science also has a cultural role, however, and presents an endless frontier to those searching for excitement. One does not have to be a Mozart, or a Shakespeare, or a Rembrandt to appreciate their genius; science can similarly be just as inspirational and beautiful. The appreciation of science, like any other facility for cultural appreciation, cannot be acquired overnight, but in view of science's vital importance to the world's economic and social welfare, it is essential that ways are found to extend radically the constituency of science. A wider appreciation of science would not only help to stem the influence of those who, perhaps unknowingly, would encourage moves towards stagnation and uniformity, but would also provide a well-informed lobby for curtailing some of the excesses of the exploiters and polluters. The constituency I have in mind would be as large in relation to the numbers of practising scientists as that of, say, music-lovers is to composers and performing musicians. It would include people who would smile and applaud when they heard a scientist, who like a musician interpreting Mozart, gave a stimulating presentation of an elegant scientific theory or observation; those who would show their displeasure when they thought someone had been wasting their time; and those in whom the spirit of exploration had been reawakened.

The problem has been recognized for a long time. In January 1961 President Dwight D. Eisenhower said in his farewell radio and television address to the American people that 'in holding scientific research and discovery in respect, as we should, we must also be alert to the equal and opposite danger that public policy could itself become the captive of a scientific-technological elite.' In reporting this remark, US Congressman George E. Brown Jr, writing in *Science* in October 1992, also said 'The fundamental challenge for us all is not to increase funding for research; it is to enhance the societal conditions that permit research to thrive.' The fact that little if any progress has been made in extending the constituency of science despite these and other influential calls is an indicator of the magnitude of the task. I believe that it is vital to find new ways to encourage the most creative of our young scientific Pelés—post-docs, lecturers, assistant professors, etc.—to devote some of their energies to sharing their inspiration with a wider audience. This would reinforce the messages from the elder statesmen, but the scientific enterprise in general must first realize that the current obsession with competition effectively discourages the young from responding to this challenge.

One of my purposes in writing this book, in addition to trying to bring about an extension to the scientific constituency, has been to help encourage those mavericks among practising scientists, or those considering joining their ranks, to follow their heretical inclinations to the full. The popular myth even among scientists, as we have discussed, is that all the easy discoveries have been made. Perhaps they have, but discoveries often appear to have been obvious only with the benefit of hindsight. If mankind has gone only a little way towards understanding this corner of the universe and the significance of our place within it, the likelihood is that we shall make substantial progress on the daunting list of problems facing the world today only when we begin to understand many things that we do not even know we do not understand. These unknown lands can be reached only by genuine explorers, who in turn need freedom to explore the wildernesses, to challenge conventions, and to ask questions that stand little chance of winning peer-preview support before they are shown to be pertinent. We need these explorers, and we need heretics in many fields of science, if only to play Devil's advocate and to test what we think we understand; perhaps also their heresy might reveal some new facets of Nature's apparently infinite variety.

Nature is a great experimenter. In the field of intelligence, Nature has evolved a species in which she has strived to endow its individuals with as much intelligence as possible—that is ourselves. The ants lie at the opposite extreme, and Nature seems to have arranged that each individual ant has the minimum of intelligence consistent with survival. The ants are one of Nature's most successful experiments: they have been around for between 50 and 100 million years; that is they were probably stepped on by the dinosaurs. Ants, however, have evolved *collective* intelligence. Ants live in colonies, but no single ant is in command, and it seems that the ants in general know instinctively what to do and who should do it. Having survived so long, they are clearly a highly resourceful species, and when Nigel Franks from the University of Bath approached us in Venture Research in 1988 with a proposal to study the factors contributing to collective intelligence, we could hardly resist.

Anyone who has watched them will be familiar with the way in which hordes of ants descend, apparently single-mindedly, on a food supply, and transport it back to the nest in conveniently sized morsels. But the observant will also have noticed that there are always one or two ants that do not follow the pheromone trail and do as the others are doing. These few dissident ants are genuine explorers, and will search for new sources of food no matter how attractive or secure the current supplies might be, or how many of their colleagues are engaged in its transport. In their collective wisdom, therefore, the ants have evolved a strategy that specifically fosters diversity and the dissidence of a few members in each colony, a strategy that would seem to have played a vital role in their survival over a very long time. We might do worse than consider the Biblical exhortation, 'Go to the ant, thou sluggard; consider her ways, and be wise.'

Short bibliography

Appleyard, Brian (1992). *Understanding the present*. Picador Books, London.

Babbage, Charles (1989). *Science and reform*. Selected works of Charles Babbage (ed. Anthony Hyman). Cambridge University Press.

Christianson, Gale E. (1984). *In the presence of the Creator: Isaac Newton and his times*. The Free Press, New York.

Clark, Ronald W. (1979). *Einstein*. Hodder and Stoughton, London.

Crick, Francis (1988). *What mad pursuit*. Weidenfeld and Nicolson, London.

Dijkstra, Edsger W. (1982). *Selected writings on computing: a personal perspective*. Springer Verlag, New York.

Dyson, Freeman (1988). *Infinite in all directions*. Penguin Books, Harmondsworth.

Harrison, Edward (1987). *Darkness at night*. Harvard University Press.

Hinsley, F.H. and others (1979). *British Intelligence in the Second World War*. (Four volumes.) Her Majesty's Stationery Office, London.

Hodges, Andrew (1983). *Alan Turing: The Enigma*. Burnett Books, London.

Kipling, Rudyard (1895–1918). *Rudyard Kipling's verse*. Inclusive Edition. Hodder and Stoughton, London.

Koestler, Arthur (1982). *The sleepwalkers*. Penguin Books, Harmondsworth.

Machiavelli, Nicoló (1981). *The Prince* (transl. George Bull). Penguin Books, Harmondsworth.

Metropolis, N., Howlett, J., and Rota, Gian-Carlo, (eds.) (1980). *A history of computing in the twentieth century*. Academic Press, London and New York.

Morita, Akio (1987). *Made in Japan*. Collins, London.

Planck, Max (1933). *Where is science going?* (with a preface by Albert Einstein). Ox Bow Press, Woodbridge, Connecticut.

Porter, Michael E. (1990). *The competitive advantage of nations*. The Macmillan Press, London.

Postman, Neil (1993). *The surrender of culture to technology*. Random House, New York.

Rosenfeld, Albert (1966). *The quintessence of Irving Langmuir*. Pergamon Press, Oxford.

Simon, Herbert, A. (1981). *The sciences of the artificial*. MIT Press, Massachusetts.

Smiles, Samuel (1881). *The life of George Stephenson*. John Murray, London.

Winterbotham, F.W. (1974). *The Ultra secret*. Weidenfeld and Nicolson, London.

Venture Researchers

The scientists listed below were participating in the first phase of the Venture Research initiative financed by the British Petroleum Company when it ended in 1990. Their current affiliations and the titles of their research programmes are also given.

Professor M. D. Bennett and Dr J. S. Heslop-Harrison, Royal Botanic Gardens, Kew, and John Innes Institute, Norwich, The nature and significance of higher-order genome structure.

Professor P. M. A. Broda, University of Manchester Institute of Science and Technology. Biodegradation of lignin.

Dr T. D. Clark, Dr H. Prance, and Dr R. Prance, University of Sussex. Macroscopic quantum phenomena.

Professor S. Clough and Dr A. J. Horsewill, University of Nottingham. Coherence in condensed matter dynamics.

Dr D. L. Cooper and Dr J. Gerratt, University of Liverpool and University of Bristol. Understanding the electronic structure of solids.

Professor A. S. G. Curtis and Professor C. D. W. Wilkinson, University of Glasgow. Understanding nerve circuits.

Dr S. G. Davies, University of Oxford. Understanding molecular architecture.

Dr P. DeKepper, Dr J. C. Roux, Dr J. Boissonade, Professor H. L. Swinney, and Professor W. Horsthemke, Centre de Récherche Paul Pascal, Bordeaux, University of Texas at Austin and Southern Methodist University, Dallas. Self-organization in non-linear chemical systems.

Professor Dr Edsger W. Dijkstra and Dr A. J. M. van Gasteren, University of Texas at Austin and Eindhoven University of Technology. The taming of complexity.

Professor Dr Edsger W. Dijkstra and Dr Lincoln A. Wallen, University of Texas at Austin and University of Oxford. The streamlining of the mathematical argument.

Professor P. P. Edwards and Dr D. E. Logan, University of Birmingham and University of Oxford. Metallic, non-metallic, and exotic states of matter.

Dr N. R. Franks, Dr J. L. Deneubourg, and Dr S. Goss, University of Bath and Université Libre de Bruxelles. Collective problem-solving.

Professor D. R. Herschbach, Harvard University. Dimensional scaling as a new calculus for electronic structure.

Professor H. E. Huppert, FRS and Professor R. S. J. Sparks, FRS, University of Cambridge, University of Bristol. Multi-phase flows in dense media.

Professor A. Keller, FRS and Professor E. D. T. Atkins, University of Bristol. Polymer transition dynamics.

Professor H. J. Kimble, California Institute of Technology. Quantum dynamics of optical systems.

J. G. Parkhouse, University of Surrey. An integrated approach to the theory of structures.

Professor A. M. Paton and Dr L. A. Glover, University of Aberdeen. Induction of novel symbioses between bacteria and higher organisms.

Professor J. B. Pendry, FRS, Imperial College of Science and Technology. Transport in disordered systems.

Professor J. C. Polanyi, FRS, University of Toronto. Surface-aligned photo-chemistry.

Professor M. Poliakoff, University of Nottingham. Supercritical fluids: an environment for reaction chemistry.

Dr A. D. M. Rayner, Dr J. R. Beeching, and Dr J. A. Pryke, University of Bath. Towards understanding multi-cellularity.

Professor I. K. Ross, University of California at Santa Barbara. Cytoplasmic control of nuclear behaviour.

Professor K. R. Seddon, Queens University, Belfast. Chemistry and physics in ionic liquids.

Professor C. H. Self, University of Newcastle upon Tyne. Biological instability.

Professor H. E. Stanley and Dr J. Teixeira, Boston University, Laboratoire Leon Brillouin (CEA–CNRS). Water in confined geometries.

Professor R. W. Tucker, Dr D. H. Hartley, and Dr D. Johnston, University of Lancaster. Geometrodynamics.

Index